Wayne Mutza

AMERICAN FIRE BOATS

The History of Waterborne Firefighting and Rescue in America

Schiffer Publishing Ltd
4880 Lower Valley Road • Atglen, PA 19310

Other Schiffer Books by Wayne Mutza:
Air 200: Aircraft of the US Bicentenntial, ISBN 978-0-7643-0388-3
Convair F-102: Delta Dagger, ISBN 978-0-7643-1062-1
Green Hornets: The History of US Air Force 20th Special Operations Squadron, ISBN 978-0-7643-2779-7

Library of Congress Control Number: 2016957414

Designed by John P. Cheek
Cover design by Molly Shields

Type set in City Medium/Lucida Std

ISBN: 978-0-7643-5271-3

Printed in China

Published by Schiffer Publishing, Ltd.
4880 Lower Valley Road
Atglen, PA 19310
Phone: (610) 593-1777; Fax: (610) 593-2002
E-mail: Info@schifferbooks.com
Web: www.schifferbooks.com

CONTENTS

PREFACE

He was an old hand at this. The dock worker wheeled his oxyacetylene cutting unit to the steel column to be cut and fitted with cargo-handling cables. The columns, too numerous to count, helped form the massive steel and timber superstructure atop the Luckenbach Steamship Co. pier at New York's Bush Terminal. The site was the nation's largest multi-tenant industrial complex, earning it the name Industrial City. Luckenbach Pier stretched a record 1,760 feet into New York Harbor from Brooklyn's 35th Street.

Undoubtedly anxious to end his work day that chilly December 3 in 1956, the worker opened the valves on his acetylene and oxygen tanks, adjusted the valve on his torch, and with a spark and slight pop, the pointy tip of his torch was lit. Pushing his welding goggles down over his eyes, he set the flame to cold metal. Satisfied when it glowed cherry red, he pressed the oxygen blast trigger, introducing high-pressure oxygen to the flame, which would chemically combine with the iron to oxidize it into molten metal, slicing the beam. Nearing 6,000 degrees Fahrenheit, the blue flame spear jetting from the worker's torch produced intense light. Transfixed, perhaps, by the brilliant glare, he paid little attention to the usual shower of sparks spraying freely about, focusing instead on the white-hot metal separating easily under his flame. He didn't notice that sparks had reached the dry fuzz of burlap bags. Nor did he hear the profound echoing of Murphy's Law—*if something can go wrong, it will.* What lay beyond his immediate work site was a recipe for disaster.

Packed into hundreds of burlap bags was more than thirteen tons of foam rubber scrap. Wind-strewn about the base of the mountain of bags were another few tons of rubber scrap that had spilled from tattered bags. The sparks swiftly transformed the flammable heap into an inferno, quickly driving off the dock workers wielding fire extinguishers. The alarm was turned in quickly, but when first-due Engine Company 228 pulled up, the blaze had already spread across the 175-foot-wide pier. It got worse. No one seemed aware that near the roaring fire lay more than eighteen tons of detonating cord, or Primacord, then deemed a minimal explosion hazard. Hard lessons were about to be learned.

Less than thirty minutes after the pile ignited, and while four-alarm units were responding, the "minimally hazardous" explosive announced its presence. The report from the New York Board of Fire Underwriters Bureau reads:

> Shortly after the fourth alarm, a terrific explosion occurred, shooting smoke and flame over 500 feet into the air. Thousands of windows were broken within a radius of a mile and serious structural damage occurred within a radius of a quarter mile. Buildings were shaken in the financial district of downtown New York two miles away and the noise of the blast was heard for 35 miles (Brooklyn Waterfront Fire and Explosion, 1957).

With land access to the pier limited due to the volume of fire, this was a job for waterborne firefighters. Engine Company 223, the fireboat *Firefighter*, had quickly slipped from its nearby berth when the first alarm came in. Nearing the scene, her pilot zeroed in on the center of the pier where flames roared from the only open doorway of the pier structure. Minutes after tying up at the pier and unleashing her 20,000 gallons-per-minute through deck monitors, the order was given to shut down and reposition shoreward to halt the fire's spread toward land.

Fortune smiled on the crew that day, for no sooner had they tied up at the new position when the blast occurred exactly where they had been minutes earlier. The firefighter securing the bow line was thrown to the deck. A firefighter aiming the monitor atop the pilothouse was blown along the roof deck until his arm caught in the chain rail, flipping him over the edge. The pilot and officer, who were thrown to the pilothouse floor, rose to see the firefighter dangling in front of the windows, all of which had their glass blown out. They rescued the hapless firefighter and quickly went into damage control mode.

The devastation wrought by the fire and explosion at Brooklyn's Luckenbach Pier on December 3, 1956, required a multiple fireboat response. At center is the large section of pier blown away shortly after the fireboat *Firefighter* moved from that point. Even greater catastrophes would tax FDNY fireboat crews in years to come. Author's Collection.

Another firefighter on the monitor deck had been blown down a stairwell, and three engineers below had been tossed about. One of the *Firefighter's* two powerful engines was knocked out of commission. On the pier, the officer of an engine crew about to take a hose line from the fireboat was blown into the water. Badly injured, he was rescued by his crew and those aboard the *Firefighter.* Farther down the pier, another engine company officer had been blown into the water, and he too was rescued. Even the *Firefighter's* quarters were severely damaged from the blast, and the watchman there was rescued by a police boat after he was injured by flying glass. Miraculously, no firefighters operating on the pier structure or aboard fireboats were killed, their salvation having come from being so close to the blast that tons of debris rocketed over their heads. Ten civilians were not so fortunate, all killed by the deadly missiles; one victim was 1,000 feet away. More than 250 people were injured. The mighty blast launched the super-heated steel of the pier's superstructure in all directions, starting fires, one of them a half-mile distant. True to their professionalism, firefighters quickly regrouped, tended the injured, and renewed their attack on the massive fire, despite the looming possibility of more explosions.

This was but one of countless incidents to which fireboats, not only of New York, but of fire departments across the country, have responded. Throughout the history of firefighting in America, such incidents were a call to action, spotlighting the glaring need for additional training of industry's workers, more prudent handling of hazardous materials, and rewriting codes to increase safety standards.

Few publications have been devoted to the fascinating subject of fireboats, the most thorough treatment having been offered by Paul Ditzel's *Fireboats* in 1989. That gap in the history of the American fire service—combined with my fondness for work boats and my fire service career—compelled me to present an in-depth look at American fireboats.

Most firefighters concede to the romance of firefighting. Theirs is a multi-faceted compulsion that may draw on childhood memories of brightly lit fire trucks with sirens wailing, or a desire to join those they admire in the noblest of professions, or simply to satisfy a curiosity or find their place in life where they can make a difference. Those who board fireboats to make a living serve in a unique maritime dimension, not only operating the most powerful fire apparatus in existence, but serving in the shadow of waterborne firefighters who forged their proud history. Their call to duty may be aboard hulking, powerful vessels that ply the massive industrial waterfronts of large cities, or smaller, modern but equally capable vessels protecting sprawling marinas. Responding to the alarm may find them extinguishing blazes that harbor the potential for catastrophic eruption, pulling survivors and bodies from aircraft, vehicles and vessels whose end is a watery grave, and everything in between.

This book is dedicated to them—and to their forefathers who skillfully and selflessly protected America's waterways and waterfronts throughout history.

ACKNOWLEDGMENTS

An in-depth treatment of the history of American fireboats, the scope of which spans more than a century and a half, is not possible without the kind and able assistance of numerous individuals, organizations, and businesses. All shared their knowledge, experience, and material, mindful of the importance of their contribution to the creation of this comprehensive study. And all shared my enthusiasm for ensuring that these fascinating vessels and their dedicated crews have a place in both maritime and fire service history.

I am most grateful to the following: John A. Ackerman; Jerry Atherton, Lake Assault Boats; Daniel Babcock, manager, Business Development, Jensen Maritime Consultants; Captain James S. Bailie, Norfolk Fire-Rescue; David Boone "Tugboat Painter"; Danny D. Bowman, chief, Fire Department of Forsyth County; Carol Borich, Minnesota Museum of Mining; Ken Buehler, Lake Superior Railroad Museum; Captain Scott Cagle, Hall County Fire Services; Michael Calamari; John M. Calderone; Chief Dave Collins, West Shore Fire District; Greg Cooper, captain, West Pierce Fire & Rescue; Paul Csizmadia, Spec3 Photography; Dede Doi; Glenn Duda; Mark Eyestone; Edward F. Flynn Jr., historian, New Haven Fire Department; Jim Flynn; Larry Ford; Ron Frehm; Bill Friedrich; Bill Fries; John M. Galla; Jan Gillespie, Humboldt Bay (Eureka) Fire; Gong Club, Inc.; Joshua Goyet, Battalion Chief, Special Operations/Marine, Virginia Beach Fire Department; Bill Hattersley; Ron Heal; Jim Hooker, The Mariner's Museum; Karl House, Puget Sound Maritime Historical Society; William G. Houston, Cincinnati Fire Department; Rick Howard; Cherie Jacobs, Media Spokesperson, Tampa Electric; Laura Jacobs, Associate Professor, Jim Dan Hill Library, University of Wisconsin-Superior; Ron Jeffers; Kent Johnson, chief, Port Clinton, Ohio Fire Department; Marvin Karsten; Brenda Keane, captain, Marine Division, Noroton, Connecticut Fire Department; Ellen Kramer, Lake Superior Marine Museum Assn.; June Larson, Door County Maritime Museum; Last Resort Fire Department; John David Lowe, CFPS, chief, Refinery Terminal Fire Company; Roger LeLeivre, KnowYourShips.com; Dennis J. Maag; Chuck Madderom; Tara Mattina, communications director, Port of Tacoma; John McClendon, president/CEO SeaArk Marine; Inc.; John McFarland; Joe Meinecke, PIO/planning manager, Tacoma Fire Department; MetalCraft Marine; Robert B. Milnes; Milwaukee Fire Historical Society; Moose Boats, Inc.; Jesse Munson, Vice President, Munson Boats; Robert Pauly, St. Louis Fire Department Museum; John L. Pennison, deputy port director, Plaquemines Port Harbor & Terminal District; Port of Houston Authority; Don Potter; Scott Price, US Coast Guard History Program; Bruce Rahlf; Emily Roantree, MetalCraft Marine, Inc.; Donald Root; David S. Rosen, Phd.; Ken Rudine; Dr. Allan Ryszka-Onions; Dennis Shell; Michael Skalley, Foss Maritime Co.; George Smith, chief, Riviera Beach VFC, Maryland; The Society of Naval Architects and Marine Engineers; Joel Stone, curator, Detroit Historical Society; Conni Spellman; Galen K. Thomaier, Seattle Fire Department historian; Bill Townsend, Tampa Fire Department; Todd Titterud; Gary Urbanowicz; Jim Van Drunen, Fort Lauderdale Fire and Safety Museum; Mike Ventimiglia; Kevin Voorhees, engineer, Humboldt Bay Fire; Frank Wegloski; Ray Wilhite; Todd A. Woofenden, editor, The Subchaser Archives; and Jack Wright, Philadelphia Fire Department historian and author.

Special mention goes to experts who willingly shared their knowledge and materials: my brother, Dale, for his time and his expertise in photography that resulted in many of the beautiful images in this book; Allied Digital Photo, Germantown, Wisconsin, for special attention to my photo needs; John A. Calderone, author and editor of *Fire Apparatus Journal*; Timothy R. Dring, expert on US Coast Guard vessels; the late Curt Elie, long-time fireboat authority whose column "Pride of the Fleet" was a regular feature in *Fire Apparatus Journal*; J. D. Floyd II, Royal Blue Ltd. Archives; Richard Nailen, fire service historian, artist, and author; Rich Schneider, West Coast firefighter, historian, and author; Britton Crosby, who manages the highly recommended site www.CapeCodFD.com; and Louis DeChime, who provided detailed insight to his involvement with the development of amphibious fireboats.

Old, rare photographs of fireboats often are difficult to obtain. Although images frequently are passed between different owners, every effort has been made to identify and credit photographers. John Landers of Brooklyn, who provided rare images of New York's early fireboats for exclusive feature in this book, has this to say:

> When I began collecting photographs of the FDNY Marine Division some forty years ago, I encountered extreme difficulty in the hobby trying to obtain photos from long-established collectors. Simply put, they were not willing to share their collections with anyone. This made no sense to me, and to this day I remain puzzled as to why some collectors won't share their photos. I made a promise to myself back then that if I was to ever obtain a significant collection, I would never deny anyone who asks or shows interest in the FDNY Marine Division an opportunity to utilize or have copies of any photo in my collection. My goal from day one was to preserve the history of the FDNY Marine Division and make sure these photos are easily available. With the help of author Wayne Mutza and fireboat researcher/photographer Britton Crosby, I believe I have kept my promise.

To Deb—for her patience, understanding, and support of my work.

INTRODUCTION

While history records the beginning of the nineteenth century as the period when "Fire Laddies" first endeavored to create fireboats, it's more likely that fires along America's waterfronts were fought from vessels decades earlier. The simple securing of land firefighting apparatus aboard vessels harks back to the mid-eighteenth century when the first hand pumpers in America arrived from England. By the end of the century, hand pumpers incorporated into small wooden boats began appearing on New York City's waterfront. As steam power was developed during the early 1800s, steam pumpers proved capable of providing continuous streams of water under pressure with far less manpower.

Volunteer firemen viewed the early smoke-belching contraptions as a threat to their existence and resisted steam technology. Although steam powerplants were elaborate and complex, the principle of steam power is simple. Water, when boiled, expands to about 1,600 times its original volume, and when confined creates tremendous pressure. That pressure is channeled to cylinders in which pistons produce mechanical power.

Eventually, steam power was refined and muscle power gave way to mechanical power. The subsequent need for horses to pull heavy steamers and the need to train men to operate machinery spelled the end of volunteer departments in major cities. The national trend to convert to organized, paid departments, however, would not gain momentum until after the Civil War, with the industrialization of port cities. Boston took the lead by establishing an organized, paid fire department in 1837, while New York held out until 1865, with neighboring Brooklyn following suit in 1869. The Chicago Fire Department was established in 1858, and in Philadelphia in 1870. Major port cities on the West Coast established paid fire departments years later—Los Angeles in 1886, and Seattle in 1883. Such was the case with major port cities along the Gulf of Mexico; Galveston, which boasted the second busiest port in the nation, established its fire department in 1885, Houston in 1895, and New Orleans in 1891.

Like the first hand pumpers in America, the first "floating steam fire engines" were developed in England, whose industrialized ports required increased fire protection. Although the designs of steam units differed according to manufacturer, their operation was basically similar. At the heart of the unit was the boiler, with water heated by coal- or oil-fed furnaces to produce steam. Steam pressure then powered pump engines. The refinement of steam powerplants spawned the more efficient steam turbine, which used a series of steam-driven blades attached to a rotor to produce power. Fireboats, both in Europe and in America, were of rudimentary design, fashioned either by securing a land steamer to a barge, or mounting a land steamer unit within a boat hull, minus the boiler. A large boiler was added and a steam-driven pump drew water through a suction port to supply hoses.

It is no coincidence that early fireboats were called "fire tugs." Mid-nineteenth-century fireboats commonly were based on steam-powered tugboats that plied America's harbors beginning in the early part of the century. Throughout these pages is evidence that the evolution of tugs parallels that of fireboats; both were proven to be indispensable elements in our marine system. Tugboats not only revolutionized shipping, but often their stalwart design served as the pattern for fireboat design. Tugboats—shortened to "tugs" in the language of mariners—often doubled as firefighting vessels, sometimes to the extent the distinction between them became clouded. Since fire aboard any vessel strikes fear into the hearts of those on the water, tug designers included a fire pump and related equipment not only for firefighting aboard, but for fighting fire on other vessels and along waterfronts. Monitors, which are fixed fire nozzles, began to appear atop pilothouses of tugs built at the end of the nineteenth century. Given their solid construction, it was not unusual for some tugs to undergo conversion to fireboats; conversely, fireboats often were converted to tugs following retirement from the fire service, extending the life of some a century, and beyond.

SCIENTIFIC AMERICAN

[Entered at the Post Office of New York, N. Y., as Second Class Matter. Copyrighted, 1891, by Munn & Co.]

A WEEKLY JOURNAL OF PRACTICAL INFORMATION, ART, SCIENCE, MECHANICS, CHEMISTRY, AND MANUFACTURES.

Vol. LXIV.—No. 10. Established 1845. | NEW YORK, MARCH 7, 1891. | $3.00 A YEAR. Weekly.

NEW FIRE BOAT FOR NEW YORK CITY—THE NEW YORKER.—[See page 148.]

Built in 1890, the powerful and much publicized fireboat *The New Yorker* served many years as the flagship of New York's fireboat fleet. While not the city's first fireboat, it was the most innovative as evidenced by steel screens along its bulwarks to protect firefighters from extreme heat. Author's Collection.

Besides leasing tugs and outfitting them with fire pumps as a cost-saving measure, another alternative was purchasing surplus military vessels for conversion to fireboat. This trend began after WWI and continued into the post-WWII era. Fire departments found that the low cost of many types of vessels, ranging from submarine chasers to small harbor craft, far outweighed the cost of building a pure fireboat from the keel up.

As auxiliary firefighting vessels, tugboats often were unavailable, even to fire departments that relied on their monitors at blazes. Nevertheless, they solved the problem of providing affordable fire protection for many fire departments and port authorities. New York City's first fireboat was the steam salvage tug *John Fuller*, which was leased from its owner on a per-call basis beginning in 1866. Before New York placed in service a purpose-built fireboat, the distinction went to Boston, which, in 1873 christened America's first steam fireboat, the *William M. Flanders,* rated at 2,500 gallons per minute, or gpm. Fireboats were identified by their name, followed, typically, by their length and, like land apparatus, were categorized by their pumping capability in gpm, a practice that continues today. Gallons per minute ratings, however, are not absolute, and fireboats, as well as land apparatus, commonly pump well beyond their rated gpm. These ratings could be misleading since the total gpm lauded by manufacturers often applied only when the engine powered pumps, and not drive systems.

A drawback of early tugs was having a limited boiler system that did not allow propulsion and pumping at the same time, a dangerous situation for crewmen when fire conditions quickly worsened. The majority of fireboats operated by major cities during the early 1800s were equipped with steam engines that used one expansion cylinder, which simplified operation but offered little horsepower. Waterfront blazes seldom were small, pointing out the need for compound engines, which doubled the use of steam, subsequently increasing horsepower and prolonging operation. The late 1800s saw the development of the triple-expansion steam engine. Among the early fireboats, only New York City's *The New Yorker* featured this engine, which was considered by many to be excessive, until shortly after beginning service it pumped continuously for five days at a major blaze. Her engines powered more than 16,000 gpm, an amazing pump rating for its day.

"Coaling" the boats was an arduous task, not only for stokers who hand-shoveled coal into furnaces, but in maintaining constant working steam pressure of at least 80 pounds to ensure rapid response; normal working pressure was between 100 and 125 pounds. Noted naval architect William C. Cowles, who designed or oversaw the construction of many of the early fireboats, professed installing the largest boiler that fit into the boat's hull. Immense boilers, which set fireboats apart from other working vessels, meant more coal. Despite the low cost and easy availability of coal, its around-the-clock use, even while the boat was berthed, accounted for the largest expenditure of fireboat operations.

Fireboats paid a premium for being in service as much as possible, unlike tugs that could "lay up" in winter or be non-operational for days at a time. The wooden hulls of fireboats had a life of between eight and ten years before major repairs were needed; their wooden superstructures lasted about four years; fireboats that operated on rivers fared even less. Before hulls were built of steel, fireboats that did double duty as ice-breakers often had steel plates fastened to their bows.

Size, too, played an important part in fireboat design. In his 1896 report, *American Fire-Boats,* engineer and professor of steam engineering at Rensselaer Polytechnic Institute Harry de Berkeley Parsons stated:

> There is little or no actual benefit to be derived from boats exceeding a certain size. Large boats become unhandy and slow in maneuvering, both of which requisites are of the greatest importance in successfully fighting a fire. When the length exceeds about one hundred and twenty-five feet, such a condition appears to be reached. Possibly the best length for rapid maneuvering around docks and slips, in connection with the displacement required for a full equipment of fire apparatus, is between one hundred and one hundred and twenty feet.

Parsons favored two smaller vessels for optimum flexibility, yet he prudently noted that, in the end, it was a paying customer's wishes that determined the size of the vessel. The deep hulls and wide beams necessary to accommodate large boilers, engines, and pumps paid dividends in stability in the water. As large fireboats began appearing at major ports at the end of the nineteenth century, observers saw little difference between their battle against fires and warships engaged in battle with the enemy. The big fireboats often were said to be of "battleship" class, a distinction easily made when observing their monitors throwing powerful streams that were capable of demolishing buildings to reach the flames.

As steam power came into its own, designers sought more efficient ways to transfer steam power to propulsion. Paddlewheels had done the job, but they came with problems and were cumbersome. Credit goes to the British for experimenting with screw propellers. The system gained little acceptance by the admiralty, however, interest in the US was strong. Successful screw propulsion built into America's first salvage tug, the *R. B. Forbes* of Boston Marine Underwriters, paved the way for other mechanically powered vessels. Twin screws drove the *Forbes* as they did Brooklyn's first fireboat, *Seth Low.* They improved maneuverability; however, as Parsons noted, "Their actual advantage appears to be limited . . . the extra floor space, weight, and

Perfection of the steam engine during the late nineteenth century led to a wide variety of fireboat designs, the most outlandish of which was proposed by inventor Sir Hiram Stevens Maxim. The 250-foot monstrosity received a great deal of publicity but not a speck of interest from the fire service. Author's Collection.

Reminiscent of images taken during the Great Flood of March 1913 that devastated fifteen states, this shows a large Midwestern fire department's makeshift fireboat using a steamer placed aboard a barge. Every river in Ohio overflowed its banks during the most widespread disaster in US history. Courtesy of the Collection of Marvin Karsten.

Intricately woven rope fenders characterized work boats during the first half of the twentieth century, such as the bow fender of USS *Quileute* (YTB-540) at Brooklyn Navy Yard in 1946. Most navy tugs were well equipped for fighting fires. Two of *Quileute's* five monitors are visible on upper decks. Courtesy of US Navy.

difficulty of management of the double machinery offset all the advantages gained." In most mechanical design there is compromise, and screw propellers compounded the condition. Wooden hulls, especially long hulls, flexed, causing leaks. Torque imposed by the long propeller shafts of screw propulsion increased the flex problem. Parsons' treatise on long hulls now proved more valid and the problem would not be solved until the hulls of vessels were made of steel.

The breakthrough in steel hull construction occurred during the 1930s when riveting gave way to welding seams, which allowed plates to be butted together, versus riveting lapped seams. Another breakthrough in marine technology during that period was invention of the Kort Nozzle by Germany's Ludwig Kort. Kort fashioned a circular shroud around screw propellers, which directed water to the screw to boost efficiency, which translated to more power; rudders were added later for sternward control.

By the time steam engines—boilers in particular—had reached peak efficiency at the end of the nineteenth century, the diesel engine was making its debut. The advantage of diesel power over steam power—less weight, less operational cost, and smaller crews—made it a natural for marine use. The first diesel-electric tug in the US was Pennsylvania Railroad's *No. 16,* which went in service in New York Harbor in 1924. The following year, Bethlehem Steel's Wilmington Yard built the first diesel-electric fireboat, the *Port of Houston.*

Despite its success and the positive features of diesel power, not all ship designers, builders, and mariners were convinced. Technically, steam offered more power since diesel engines lost some of their power in transmission through gearing to screws, while steam engines, with slower turning speeds, had direct drive. The major drawback of steam power was boiler explosions. If not carefully monitored and maintained, steam boilers could over-pressure and explode. It was estimated that during the steam era, boiler explosions accounted for more than twenty percent of the mishaps involving steam-powered vessels. The debate over steam versus diesel power lasted well into the 1930s, when diesel engines were refined. They gained the acceptance of even the most stalwart mariners when they proved feasible by being safer, more efficient, requiring less fuel, providing power faster, and requiring less space. The latter proved a boon to fireboat designers who filled the spaces once occupied by massive boilers with multiple engines, fire pumps, and various types of equipment. With improvements in reduction gears, diesel power came into its own during the 1940s.

Gasoline engines too had made an impact; however, their use in fireboats was limited due to the explosive nature of their vapors. By 1930, the spectacular era of steam drew to a close as many steam-powered fireboats were overhauled and converted to diesel or diesel-electric power. Some fire departments, including New York's, remained wary of diesels, opting to convert from steam to gasoline engines. Conversions using either system usually resulted in significant increases in gpm ratings. Seattle's *Duwamish,* for example, built in 1909, forty years later, was converted from steam to diesel-electric power. The conversion increased the gpm rating of the 120-foot battleship-class fireboat from 9,000 gpm to 22,800 gpm; *Duwamish* was then billed as the world's most powerful fireboat. The City of Portland, Maine, helped set the standard for future fireboat propulsion when it commissioned its ninety-foot *City of Portland* in December 1931. As the nation's second diesel-powered fireboat, *City of Portland* featured five diesel engines, which, through direct connection to four fire pumps and reduction gearing to the propeller shaft, allowed power to be shifted from propulsion to pumping as needed.

Industrial growth during the late nineteenth century meant expanding waterfronts across the nation, sometimes built at a rate with fire potential far exceeding the capabilities of metropolitan fire departments. Major port cities that suffered waterfront conflagrations about the turn of the century include New York, Seattle, Boston, Brooklyn, Milwaukee, Chicago, San Francisco, and Baltimore. The handwriting was on the wall for city leaders—fireboats were needed. Busy, recklessly crowded ports, whose wharves, piers, and warehouses were built mainly of wood and without fire codes, in some cities accounted for nearly half of major fires. Insufficient numbers of land companies led to spreading blazes, and city water supplies often were inadequate. Fireboats not only provided sweeping coverage of conflagrations, their water supply was unlimited and they were capable of throwing powerful streams hundreds of feet.

From early on, their ability was not limited to waterfronts; hose could be laid inland for great distances, thanks to horse-drawn—and later, motorized—"boat wagons," or "hose wagons," whose sole function was to carry large diameter hose. Often mounted to such apparatus were large steam monitors similar to those of fireboats, and referred to as "deck pipes" or "high-pressure turrets." Key among the many features built into fireboats was multiple discharge outlets to supply these water cannons, along with lower pressure hose lines for firefighters on shore. In view of their tremendous pumping capability, and to maximize use of their fireboats as

Using a splicing fid, a Los Angeles firefighter works with large-diameter rope. A fid traditionally was made of whalebone or wood to hold open rope or canvas knots, or to part the lays to splice rope. Today's fids are hollow metal, but still spike-shaped. Author's Collection.

pumping stations to bolster or replace municipal water systems, many cities laid large, high-pressure water mains with fireboat connections. In Cleveland in 1888, a permanent, 6-inch-diameter main was laid underground from the Cuyahoga River. Milwaukee in 1893 laid three miles of underground fireboat pipelines; Detroit and Buffalo followed. Boston in 1897 began construction of a 5,000-foot main leading directly to the fireboat's berth. A report explained:

> As the fireboat is always under steam, it was the work of but a moment on the receipt of an alarm to make connection with the mains and start the pumps, the firemen in the meantime having attached their hose to the hydrants at which a pressure of 200 pounds could be secured (Firemen 1901).

Brooklyn and New York installed similar systems, vowing to use them only as a last resort due to saltwater's corrosive effects on pipes and equipment. Philadelphia is credited with the first large-scale high-pressure system installed in 1904. The concept caught on, and with the added benefit of reduced insurance rates, other East Coast cities, including Washington, Baltimore, Atlantic City, Newark, Providence, and Hartford, planned installations of high-pressure water systems during the early 1900s.

The city of San Francisco lived up to its reputation as "the city that knows how," when, in the aftermath of the earthquake and conflagration that devastated the city in 1906, city officials were quick to plan improvements for fire protection. Completely independent of the domestic water supply and built solely for firefighting, an elaborate system of water mains was installed, beginning in 1909. Using nineteenth-century hydrants and cisterns, plus pumping stations and zone controls, the system could be augmented by a pair of 120-foot, 9,000-gpm fireboats, which were ordered specifically as components of the system. San Francisco's two modern fireboats, having a combined capability of 34,000 gpm, like their predecessors, can augment the system from the bay.

Since it proved more economical to build ships where the engine and other machinery were built, by the turn of the century the majority of East Coast shipbuilding took place along the northern Atlantic coast in cities such as New York, Baltimore, and Philadelphia. Leaders of municipalities, usually with the aid of a naval architect or marine engineer, consider a number of factors when designing a fireboat: size, which affects berthing facilities, staffing and maintenance, and fuel costs; purchase cost, whether new from the manufacturer or conversion of a used vessel; construction, which usually depends on climate, water depth and salt or fresh water; and special features, such as monitor towers necessary for reach, and foam systems, which are vital in areas having petroleum products.

A good example of using proximate resources in the design and construction of a fireboat is Milwaukee's *Deluge*. Her Midwestern composition is obvious, having been designed in Chicago, built in Michigan, and her power plants and pumps built in Milwaukee.

What's in a name? Fireboats usually are named before their construction is complete and the name is applied to a builder's plate mounted to the pilothouse. Name boards are fastened to the pilothouse and the name was painted or carved onto a trailboard on the bow. Typically the name honors city mayors, civic leaders, officials who were instrumental in the betterment of fire service, fire chiefs, and firefighters killed in the line of duty. When a vessel's keel is laid, shipbuilders assign a hull number, and the boat is given a mandatory five- or six-digit US registration number. In marine tradition, launching the vessel is done with much fanfare, which includes dignitaries, a band, and, of course, those who built the vessel. The all-important christening is usually performed by a woman. The same tradition holds that vessels are affectionately and respectfully referred to in the female gender. When a vessel is launched, or commissioned, there remains much work to be done as she is merely deemed floatable and has yet to be outfitted and undergo rigorous testing for service. The launch itself is accomplished through a time-honored sliding down the ways using two primary methods; dramatic side launch or stern-first into the water, with the former preferred over an astern slide down the ways. For maintenance of vessels, a powerful crane could be used to bring the boat out of the water, or various types of dry docks employed. A graving dock functions much like a canal lock, first filled with water to allow the vessel's positioning for work, and then pumped out, leaving the vessel to rest on keel blocks and supported by timber called "strongbacks" wedged against the hull. A less expensive, easier method, especially for smaller boats, used a marine railway, which allowed the boat to be pulled up onto an embankment.

Adding to the size of some boats were living quarters, since cities often did not provide dockside quarters for crews. Such arrangements gave the boats more latitude in positioning at sites where they were most needed, making communication vital. Communication between fireboats and shore, and between boats themselves, initially was carried out with steam whistles or semaphore, the latter a visual system using signal flags or lights. During the late 1800s, fireboat quarters were alerted in cities that installed box alarm systems. The wireless telegraph, using Morse code, was applied to marine use at the beginning of the twentieth century, eventually giving way to wireless radio.

Overhauling and re-powering early fireboats with diesel and diesel-electric engines and centrifugal pumps during the era of motorization changed their appearance dramatically. Gone were towering funnels, bulky coal storage bins on decks, and ornate trappings. Hulls could be re-painted traditional red to replace dark colors that masked the dirty job of dumping huge amounts of ash left by coal-burning steam boilers. The fenders that characterize work boats and speak volumes about their close-in work usually were made of oak in the early days of shipping. These became expensive and broke frequently, so they were replaced by rope fenders made by weaving old tow lines. Bow fenders were rope mats through which short strands were tied and pulled apart to form whiskers, which protected the fender itself. Eventually, rubber tires comprised the fender core and were covered with a rope mat. Rubber tires alone were later used, with the production of molded rubber fenders begun during the 1930s. Crafting rope fenders became a dying art, along with creation of the ornate features that identified the steam era. Handsomely appointed pilothouses built with select woods disappeared when steel was used to construct the superstructures of vessels.

In keeping with its pledge to protect US waterways, the US Coast Guard just prior to WWII began placing the first military fireboats in service, eventually forming the largest fireboat fleet in US history. During the war, the value of also outfitting army and navy vessels for firefighting would become apparent. During postwar years, a number of these vessels were converted to full-fledged fireboats by municipal fire departments in San Diego, Baltimore, Jacksonville, Stockton, Newark, Boston, and Savannah, to name a few cities. A later chapter examines the firefighting service of US military vessels, and we take a fascinating look at military amphibians-turned-fireboat.

As marine technology improved, fireboats were designed more specifically to adapt to their response area. Appearing in some Midwestern cities were low-profile boats that could pass beneath street-level bridges. Other features of fireboats that worked in severe winter climes included reinforced bows for ice-breaking, and hot water jackets surrounding sea chests to prevent icing of water intakes. Fireboats that protected sprawling petroleum and natural gas facilities were equipped with foam systems. Across the nation the practice of building dockside crew quarters eliminated the need for large deckhouses, thus allowing more spacious trunk decks. The all-important purchase and operating costs encouraged the design of smaller vessels, many of which were built from commercial boats. Battleship-class fireboats would have their place along sweeping waterfronts of major cities, with smaller, faster, more maneuverable vessels added to the fleet. The city of Chicago not only championed the use of small boats, its use of a waterjet drive at the beginning of the 1960s helped revolutionize this form of marine propulsion.

Although the invention of the waterjet in 1953 is credited to New Zealander William Hamilton, during the 1930s fire apparatus engineer Keenan Hanley perfected a pump that supplied monitors and provided water-jet propulsion. In Indianapolis in 1959, Indiana Gearworks president John Buehler partnered with Hamilton to secure manufacturing rights to create the waterjet unit. Buehler then combined his knowledge of fiberglass with the purchase of sailboat maker Heckel Plastics Products Company. Buehler Turbocraft was formed as the marine division of Indiana Gearworks and began turning out jet boats that ranged from sixteen to twenty-two feet, and 109 to 230 horsepower, respectively. To broaden market appeal, Buehler Turbocraft designed a line of patrol and rescue craft.

The heart of the Buehler boat was the multi-stage, axial flow Hamilton Marine Jet Unit. Water was drawn in through a port in the boat's keel, with thrust achieved by discharging water rearward. Steering was accomplished by deflecting the high-velocity stream port or starboard, or reverse. The high-pressure flow could be diverted to monitors, usually at 150 psi, through a three-inch-diameter fire main. The axial flow turbine was powered by an Interceptor inboard engine supplied by Dearborn Marine Engines, of Madison Heights, Michigan. The engine normally was mounted amidships, with the fire pump mounted forward and the propulsion jet pump situated rearward of the engine. This arrangement allowed the boat to simultaneously pump water and maneuver.

Bolstered by the debut of aluminum and fiberglass in boat construction during the 1950s, small jet-powered boats became instantly popular, with the market for high-powered runabouts peaking during the 1970s. Although they offered increased reliability and less maintenance than prop-powered boats, jet-powered boats were considerably less fuel efficient, which led to their decline in popularity during the 1980s fuel shortages. Regardless, jet boat technology has come a long way in furthering marine propulsion, with many units relying on diesel power, and eventually diesel-turbines. Also appealing is the waterjet's ability to counter the powerful back-pressure of streams to keep the boat on station; one or more monitors of older fireboats usually were committed to this task. The absence of underwater appendages such as screws and rudders not only allows shallow water operations, jet boats can operate safely around people in the water. Improvements in waterjet propulsion systems enhanced maneuverability at slow speeds, and they have been adapted for steering thrusters, which counteract reaction from high pressure monitors.

The 1950s signaled a decline in fireboat activity, which is attributed to several factors. More stringent fire codes and enforcement, along with fire resistant structures built to replace those made of wood, made industrial ports safer. As cities modernized waterfront facilities, fire departments upgraded their land apparatus and tactics for more effective operations.

Fires aboard ships proved challenging for fireboat crews, especially if boarding was necessary. The immensity of this arson fire aboard the *Seawise University* ruled out boarding. Formerly the RMS *Queen Elizabeth*, *Seawise University* burned after sailing from Florida en route to Hong Kong for outfitting in January 1972. Author's Collection.

Often, industry moved inland, with containerized cargo fast becoming the ideal method for transporting cargo both aboard ship and on land. Improvements in ship design also played an important part in reducing marine fire hazards. A 1973 report by the US Maritime Administration was one of the first steps in establishing standards in fireboat construction. The report called for fireboats to be capable of being operated by a two-person crew, have a twenty-year hull life, be capable of pumping 5,500 gpm for eight hours, have a dash speed of 30 knots, and be equipped with multiple monitors. Shipbuilders and port authorities are held to higher standards of safety, with oversight by agencies such as the US Coast Guard and the American Bureau of Shipping. Both agencies have published extensive guidelines for marine fire-fighting systems, along with the National Fire Protection Association, which publishes *NFPA 1925 Standard on Marine Fire-Fighting Vessels,* which details requirements for the construction of fire-fighting vessels, the conversion of existing vessels, and testing and maintenance. The International Convention for the Safety of Life at Sea (SOLAS), first formed in 1914 as a result of the RMS *Titanic* disaster, ensures that nearly all passenger, cargo, and tanker ships worldwide comply with minimum safety standards in construction, equipment, and operation. Despite guidelines, the design of fireboats is determined largely by the needs of a community, ensuring their individuality.

Replacing much of the industrialization along the waterfronts of US cities are development projects comprising expansive marinas, tourist districts, housing, and commercial complexes. Such high-density developments have changed the face of waterborne firefighting. As more and more people are drawn to the water, fire departments and port authorities face new challenges, paramount among them life hazard. As large fireboats were retired, they were replaced by smaller vessels, although technology enables them to pack plenty of punch to fight fire. Smaller, yet growing, waterfront communities too found it necessary to invest in fireboats. Like the fire departments they serve, which have become more multi-purpose, modern fireboats are more aptly labeled "fire rescue" boats. Their versatility helps to justify the expense. Besides pumping water, they can serve as dive-rescue platforms and command posts. In keeping with the modern trend of doing more with less, many firefighters trained as emergency medical technicians or paramedics serve as crew aboard vessels that also serve as EMS vehicles that bridge waterways. Budget constraints and subsequent political pressure that calls cost-versus-demand into question has prompted some fire departments to resort to small boats that require fewer crew. Advanced designs, along with lightweight metals and composite materials, make these vessels highly maneuverable, which proves beneficial in crowded marinas where they are less likely to create wakes that damage other boats.

With refinement over the years, waterjets emerged as the preferred method of propulsion for modern fireboats. MetalCraft Marine of Ontario, Canada, became a leader in the industry by combining its Kingston jet boat hull with HamiltonJet power units. The company's wide range of Firestorm models uses Hamilton waterjets and Cummins diesel engines. MetalCraft's aluminum hull, which was introduced in 1987, provides maximum lift and low trim angles that permit heavy gross weights. Another company trademark is a "push knee" incorporated into the bow.

Having earned prominence in the workboat industry through family ownership, SeaArk Marine of Monticello, Arkansas, also relies on Hamilton waterjet propulsion with Cummins diesel power. SeaArk, like other builders, also produces workboats powered by gasoline outboards, which can be combined with waterjets.

Paramount in fireboat construction is propulsion, which has come a long way since the days of steam power. Although the conventional diesel-powered, fixed-pitch screw, or propeller, has stood the test of time for overall efficiency, systems other than waterjets were found to be equally effective. Credit for breaking new ground in fireboat propulsion goes to Washington's Tacoma Fire Department for becoming the first in the US to employ hovercraft technology. This technology, technically labeled Surface Effect System (SES), was nothing new, having undergone study by the US Navy beginning in 1960 for patrol craft. The concept of a large, catamaran-hulled vessel skimming the surface of the water on a cushion of air was a radical departure from the century-old concept of diesel-powered, sturdy steel ships that had evolved from heavy-duty tugboats. But the deviation from tradition didn't end there—Tacoma's unique fireboat design sparked further controversy; much like the design of some high-performance aircraft, Tacoma's fireboat design was to be based on performance, with specifications worked out later. Tacoma deputy chief Tony Mitchell explained: "Because the product we were pursuing was unique and nonexistent, and because we had limited resources which did not include the development cost, we opted for primarily a performance specification."

Although US shipbuilders shied from the project when Tacoma solicited bids, controversy flared anew when Tacoma signed with the only bidder, Great Britain's Vosper Hovermarine. In the eyes of traditionalists, not only was Tacoma's break from the practice of using local or regional shipbuilders adding insult to injury, the fireboat would be painted lime green, not traditional red. Tacoma ordered two SES fireboats, which arrived during 1982. Initial trials with the unique pair proved that they met or exceeded all expectations; the skeptics were stilled and fireboat experts came to Tacoma to see the new look in fireboats.

The Fire Department of New York during the early 1990s purchased twin Surface Effect Ships; however, the FDNY would not experience the level of success near that attained by the Tacoma Fire Department.

Unconventional propulsion systems were slow to gain acceptance in fireboat design. Despite the long history of Z-Drive and Voith-Schneider cycloidal drive systems, and their growing popularity in workboats, few would be found in fireboats. Both systems combine propulsion and steering functions in a single unit.

The wave of the future in propulsion for fireboats is called the Voith-Schneider Propulsion System, or VSP. Projecting downward from the forward hull is a group of lengthy rotating blades providing superior thrust and maneuverability. Aft of the unit is a large skeg, or fin, to aid directional stability. Author's Collection.

Although unfamiliar to US marine operations until the 1990s, the Voith-Schneider drive system traces its history back to Germany in the late 1920s. Austrian inventor Ernst Schneider teamed with Voith engineers to fashion his water turbine into a combination propeller-rudder. This highly maneuverable system uses a circular array of hydrofoil-shape blades that extend vertically downward from a circular plate in the hull. Each blade—which is between six and eight feet long—rotates on its own axis, its pitch in sync with the plate's rotation with control similar to that of helicopter rotor systems. The result is almost instantaneous thrust and superior maneuverability. In 2003 Los Angeles placed in service America's first fireboat with cycloidal drive, the 105-foot, 38,000-gpm *Warner L. Lawrence,* which features two Voith-Schneider systems.

The Z-Drive system, also called an azimuth thruster, was the invention of Germany's Joseph Becker in 1950. As the system was refined, it quickly became popular for workboats. The Z-Drive, which takes its name from the letter "Z" formed by the cross section of its configuration, features a shrouded propeller that rotates about the vertical axis to direct thrust, eliminating the need for a rudder. Azimuth drive, generally, is positioned below the stern. Although less efficient and slower to react to control input than the cycloidal system, azimuth drive is less expensive. Of the two systems, the cycloidal is least desirable for operations in shallow waters because its drive components extend well below the hull. The system, along with a large stern skeg, or fin, to aid directional stability, gives

Los Angeles' *Lawrence* an extreme draft of fourteen-and-a-half feet.

Fire officials found it increasingly difficult to justify having large fireboats. The number of calls had declined, and vessels were aging and becoming more expensive to overhaul. The operation and staffing of large vessels typically presents as a fat target for the budget axe. Due to budget constraints, some cities found it necessary to terminate fireboat service, which shifted the responsibility for marine fire and rescue service to other authorities or commercial interests. The rise in the number of law enforcement patrol craft often filled the gap, particularly since 9/11, when record amounts of federal funding to counter terrorism became available. That same funding, in many cases, allowed fire departments to purchase modern fire-rescue vessels. This proved beneficial to the US Coast Guard (USCG), whose responsibilities have increased significantly since 9/11. Despite its increased workload, the Coast Guard has maintained a long-standing policy of "assistance as available." Although the Coast Guard has not operated fireboats since WWII, the service traditionally assists local authorities that are principally responsible for firefighting and rescue at US ports and harbors. The Coast Guard is a vital resource to fire departments that work closely with USCG Captains of the Port, whose authority was markedly increased with the Ports and Waterways Safety Act of 1972.

Although no specific licensing or certification requirements exist for fireboat crew members, most fireboat captains are licensed and certified through the Coast Guard. In some jurisdictions, all crew members are required to be USCG-licensed. Recognizing the need for teaching fireboat crew members a host of special skills that go beyond those associated with traditional land firefighting, many fire departments establish training and certification programs.

Throughout the history of the American fire service, fireboats have evolved from hand-operated pumps aboard wooden barges to stout, steam-powered behemoths and mighty battleship-class ships with diesel powerplants. Their amazing transformation has moved from floating pumping stations during major disasters to modern, sophisticated vessels brimming with advanced technological systems that allow them to operate in the most hazardous environs. Their long, tireless service and contribution to the nation's history went largely unheralded until 1989, when six fireboats built during the first half of the twentieth century were designated National Historic Landmarks. They are: New Orleans' *Deluge,* Seattle's *Duwamish,* Tacoma's *Fireboat No. 1,* FDNY's *Firefighter,* Los Angeles' *Scott,* and Oakland's *City of Oakland/USS Hoga;* Buffalo's *Cotter* was added in 1996, and the FDNY's *Harvey* in 2000. Read on and discover how fireboats have adapted over time to protect and serve in US waterways in the true spirit of America's dedicated, skilled firefighters.

Since the history of fireboats in America spans nearly a century and a half and includes a wide range of vessels, there is no precise method to present their development and operation. I chose a regional presentation with which readers can identify. A number of these unique vessels thought to have been lost to history are presented in these pages. Included is a section devoted to military boats' often overlooked contribution to US history, along with coverage of support equipment that has proven vital to fireboat operations.

After serving long and distinguished careers, many fireboats begin new careers, while others, such as Chicago's *Medill* seen here, were relegated to storage on land to be transformed by the elements into artful studies. Courtesy of Charles Pomazal.

Against a backdrop that would hold enduring significance for the FDNY and for the nation, the fireboat *John D. McKean* dominates a waterfront blaze in Manhattan. Courtesy of Ron Frehm.

CHAPTER 1: GUNS AND HOSES

Fighting fire does not immediately come to mind when one thinks of military watercraft. We think of deck guns on vessels as being formidable weapons for naval warfare, not as fire monitors to combat another enemy—fire. But given the size of the nation's naval fleets, it is no surprise that fireboats and boats with firefighting capability were among thousands of army, navy and Coast Guard vessels. Seldom, however, has their monumental contribution to the fire service in America been recognized.

The nation's first armed naval force, called the Revenue Marine Service, is as old as the constitutional government, with its first vessel having begun patrols on the Delaware River from the Port of Philadelphia in 1789. The service evolved into the Revenue Cutter Service and in 1915, when merged with the US Life Saving Service, became the US Coast Guard. An important function of the service was enforcement of an 1838 Congressional act requiring that steam vessels be equipped with fire pumps, hose, and other safety equipment; boilers were to be inspected every six months. By the late 1800s, tugboats had become part of the Coast Guard's fleet. Besides their primary duties, which included ice-breaking, law enforcement, and search and rescue, all were capable of fighting fires. The common habitat for Coast Guard tugs was protected waters, stationed at major port cities. Although tugs would prove indispensable in commercial shipping, in their infancy during the early 1880s they were used mainly for naval shipping. Navy ships usually were the largest afloat and their maneuverability with the aid of tugs in wind and tides gave them a tactical advantage. Fire pumps installed in naval ships for self-protection soon included discharge hose manifolds and monitors on tugs, allowing them to double as fireboats.

The earliest recorded use of a military fireboat cites the navy's USS *Rescue* built in 1861 as a gunboat for the Union Navy. After the Civil War, she was converted to a fire tug to serve as then-City of Washington's fireboat. Unique among the US Navy's fleet at the turn of the century was a purpose-built fire tug named *Leslie (YFB-7).* Although YFB was the designation for ferry boat or launch, *Leslie's* crew

The US Navy fire tug *Leslie* (YFB-7) mounted a single monitor, which was atop its boot heel pilothouse. Skylights for light and ventilation were standard fixtures of a long deckhouse. The *Leslie* is seen here at US Navy Yard Mare Island in September 1903. Courtesy of US Navy.

preferred Yard Fire Boat. The 75-foot wooden vessel was built in 1902 as a yard craft at Mare Island Naval Shipyard, Vallejo, California, and served her entire navy career as Mare Island's fireboat. During the 1906 earthquake and fire that devastated San Francisco, from April 18–23, under command of US Navy Lieutenant Frederick N. Freeman, the Torpedo Boat Destroyer *Preble,* along with the *Leslie* and yard tug *Active*, distinguished themselves by supplying water to San Francisco firefighters and fighting major blazes. The 107-foot *Active* also shuttled navy firefighters between Mare Island and San Francisco.

Coast Guard supervision of vessel activity in US harbors began with its inception in 1915 and took on even greater significance with the Espionage Act of 1917. Since most of the nation's shipments of munitions sailed from New York during WWI, the city required the largest USCG port security presence, which included nine harbor cutters and tugs.

The first military vessels known to have been acquired by civilian fire departments for conversion to fireboats were US Navy sub chasers built during WWI. Two wooden, 110-foot sub chasers went to Eastern Seaboard fire departments—*SC 145* to Jacksonville and *SC 428* to Baltimore. During the same period, Baltimore acquired an army 60-foot tender, and two ex-military tugs went to Norfolk, Virginia, for fireboat duty.

Strongbacks and a gangplank help support the *Leslie* while in drydock at Mare Island in 1914. Vessels were floated into the massive concrete dry dock and the water pumped out. The process was reversed when repairs were completed. Courtesy of US Navy.

The *Leslie* docked at Building Ways 2 at Mare Island, California, in 1921. The battleship *California* is in the background, and at right are two 65-foot yard motor tugs. Courtesy of US Navy.

World War II

A major blaze at a US waterfront during WWII often drew a mix of marine firefighting vessels from various agencies. With the outbreak of war imminent in 1939, US Coast Guard responsibility for safeguarding harbors would increase dramatically with President Roosevelt's proclamation of neutrality. Additional presidential mandates the following year had the Coast Guard assuming functions of other government agencies to clarify jurisdictional boundaries at all levels of government. When relentless attacks by German U-boats on Allied shipping signaled that the war had arrived at America's Atlantic coast, concern and even paranoia over the likelihood of sabotage of war materiel was pervasive. Port security was deemed vital and included fire prevention and fighting fires. Besides sabotage, the dangers of smoking and of cutting and welding at repair facilities needed to be addressed.

This was the last of twenty-four requisitioned commercial or private vessels with known hull numbers 7701 through 7723 converted to Coast Guard fireboats. Under the 1943 re-designation system, this became number 64018-F. Six 500 gpm pumps on deck delivered 3,000 gpm to three monitors on a raised foredeck. Vessels in this class normally had one bow monitor. Courtesy of US Coast Guard.

The burning of the French liner *Normandie* in New York Harbor in January 1942 pointed out that lines of authority remained blurred, prompting President Roosevelt on February 25 to issue Executive Order 9074 directing the Coast Guard to impose strict safeguards at US and territorial waterfronts. With shipments through major US ports vital to the war effort at home and overseas, President Roosevelt expanded the mission of Coast Guard captains at twenty-nine strategic ports, only twenty-three of which had marine fire protection. With dangerous cargoes, such as ammunition and fuel, concentrated at berthing areas, fire inspections were conducted regularly and guards protected wharves; a no-smoking policy was enforced, and charged hose lines were at the ready during ship repairs that required "hot work," such as cutting and welding. Within eighteen months

Even before embarking on a long career as a municipal fireboat, the navy tug USS *Hoga* distinguished itself during the attack on Pearl Harbor on 7 December 1941. Here, its crew makes a valiant effort to save the burning USS *West Virginia*. Courtesy of US Navy.

The US Coast Guard ensign flies high on this Hanley fireboat of the USCG Port Security Patrol. More than 100 of the 30-foot, 2,000-gpm boats were built during 1942 and 1943, remaining in service until 1949. Courtesy of US Coast Guard.

of the order, Coast Guard fireboats had responded to nearly 3,000 waterfront fires across the nation; by mid-1944, Coast Guard fireboats were operating at 121 ports.

Coast Guard firefighting and prevention operations were tied closely to municipal fire departments, whose members were grateful, having found themselves hard pressed to handle the wartime increase in port activity. Fire departments were strained by manpower depletion when firefighters went off to war, and by government restrictions on new fire apparatus when the production of war materiel became top priority. The Coast Guard could offer little help until it added significant manpower and equipment to meet the demands of its expansive wartime duties. Some relief came in the form of temporary Coast Guard reservists who manned fireboats in Portland, Maine; Providence, Rhode Island; Washington, DC; Cleveland, Ohio; and Galveston, Texas. Units at Memphis, Tennessee; Rock Island, Illinois; and St. Paul, Minnesota, were manned by Coast Guard reservists.

An example of the Coast Guard commitment to waterfront protection is the Pacific Northwest, where Seattle's 10,000-gpm *Duwamish* was manned by Coast Guardsmen, with thirteen types of fireboats and barges positioned strategically along Seattle's waterfront. Smaller types built at the war's onset from discarded Lundeen merchant lifeboats were stationed at Portland, Oregon, and at seventeen ports along

US Coast Guard officials, including Commandant Admiral Russel R. Waesche, at right, observe testing of the second-built Hanley fireboat (30033-F) at Washington, DC. Hanley boats were numbered 30032 through 30131. Courtesy of US Coast Guard.

the Pacific Northwest. The Lundeen was a metal, screw-propelled, 30-foot vessel with a 500-gpm pump installed amidships. Although its screw was manually driven, the boat was usually towed to the fire. Coast Guardsmen found the "Lundeen Firebarge" extremely effective for under-pier fires and in tight places where large boats could not maneuver.

Two types of specially built fire barges, both rated at 4,000 gpm, proved underpowered and unstable in rough current but packed a powerful punch when all monitors were used. During 1942 and 1943 fourteen 50-foot fire barges were built by six shipbuilders—all in the state of Washington, with the exception of Wisconsin's Peterson Builders. Just over 19 feet in width, these carvel-built barges had a top speed of 8 mph and were stationed on the Great Lakes. They wore numbers CGB-50300-F through 50313-F. Eleven 60-foot fire barges, seven of which were built by the Walter W. Johnson Company at Oakland, California, in 1943, were assigned to Pacific ports until 1946. Nine crewmen manned each of the plywood barges, with those built by Johnson identified as CGB-60012-F and 60013-F, and 60015-F through 60019-F. Both types of fire barges were powered by 286 hp Chrysler diesel engines and mounted eight Chrysler 500-gpm pumps, in addition to equipment for foam operations.

There seemed to be no limit to the types of vessels the Coast Guard would employ to accomplish their goals. That especially held true when USCG officials devised a plan to augment city fireboat fleets. They worked quickly to acquire more than 150 private and commercial trade vessels for conversion to fireboats. Under the War Powers Act, which allowed acquisition of civilian property for government use, vessels were sold or temporarily assigned to the Coast Guard and designated Coast Guard Reserve (CGR) or Coast Guard Auxiliary (CGA). Many of these boats were manned by their owners who signed on as temporary reservists. Boats of various types that measured from 30 to 96 feet, from fishing trawlers to pleasure craft, were given a coat of gray (or red) paint, light armament, hose, and as many P-500 fire pumps as decks could accommodate. The Chrysler P-500 was a portable, centrifugal pump driven by a two-cycle, four-cylinder, water-cooled gasoline engine.

A typical example of a commercial vessel pressed into wartime service was the 70-foot fish tug *Tessler,* built in 1905 by Johnston Brothers of Ferrysburg, Michigan. The steel-hull, 57-gross-ton *Tessler* was owned by a Kaszube family, which was part of the fishing community that inhabited Milwaukee's Jones Island. In May 1942 the Tessler family sold the rugged boat to the government, which turned it over to the Coast Guard for conversion to fireboat *CG-70004-F. It* was assigned to the 9th Naval District for duty at South Chicago until September 1945, and in 1946 was sold as military surplus to Pelican Towing Co.

Twenty nine Wheeler fireboats were built for the Coast Guard during 1944 and 1945. They were piloted from a wheel mounted at the aft end of the deckhouse. A unique pipe system surrounding the deckhouse sprayed a protective curtain. A harness that incorporated straps riveted to the hull allowed hoisting the boat for maintenance or placement in storage cradles at fire station berths. Courtesy of US Coast Guard.

Believed to be a Hanley boat, this fireboat mounts dual monitors geared for elevation and traverse, along with gated discharge outlets for hose connections. Courtesy of US Coast Guard.

The roomy afterdecks of these tugs converted to USCG harbor tugs allowed the installation of four 500 gpm pumps. At left is the former tug *Joseph Alioto* alongside CG-72004-F, formerly the tug *El Gabilan*. Both were assigned to the CG 12th Naval District San Francisco when they fought this blaze in 1944. Masts were non-functional as sail rigs, but bore underway tow lights and antennae. Life rafts commonly were lashed to rigging to save space. Courtesy of US Coast Guard.

Groups of small firefighting vessels distributed to harbor patrol divisions included 30-foot, steel-hull Hanley fireboats, 103 of which were ordered by the Coast Guard. Equipped with four 500-gpm fire pumps, the Hanley boat was the brainchild of engineer Keenan Hanley, who sought a solution to fighting fires when flooding rivers during the late 1930s blocked land apparatus. Originally an engineer with Ahrens-Fox, Hanley in 1928 joined fire apparatus builder Prospect Fire Engine Co. in Ohio. In 1930 he assumed ownership of the firm and later formed Hanley Engineering Services, where he devised a pump that not only supplied firefighting appliances but propelled the boat using a 3-inch-diameter waterjet port in the hull. Hanley's first boat went to Wheeling, West Virginia, in 1937, with more cities purchasing boats before the Coast Guard placed its large order at the war's onset.

The influx of boats to harbor patrol divisions didn't end with the Hanley boats; twenty-nine 40-foot fireboats, each with seven monitors, were built by Brooklyn, New York's Wheeler Shipyard for assignment along the East Coast. Wearing numbers CG-40335-F through 40364-F, these wooden fireboats carried a crew of ten and were powered by a 143 hp Chrysler gasoline engine. Six 36-foot, 2,000-gpm fireboats built in Fulton, Texas, by Perry E. Bass Boat Works were assigned along the Gulf of Mexico coastline; these boats were numbered CG-36024-F through 36029-F. By 1944, nearly 300 Coast Guard fireboats were serving the nation's ports. Due to their lack of speed, most were placed at strategic harbor areas. Augmenting the fleet were fifteen 64-foot, wood-hull, diesel-powered harbor tugs built during 1943 and 1944. The largest number of newly constructed Coast Guard fireboats was assigned to the 13th Naval District of the Pacific Northwest, while the second largest number went to the Great Lakes 9th Naval District. The 64-footers, along with seventeen 110-foot harbor tugs built between 1934 and 1944, had firefighting capability and performed ice-breaking duty. Even picket and small patrol boats were outfitted with small P-60 fire pumps, called hand-billys. Developed in 1938, the P-60 was a rotary-type pump driven by a two-cycle, two-cylinder gasoline engine. The 106-pound unit was rated at 60 gpm at 100 psi and had the added feature of a proportioner for producing mechanical foam. Many fire monitors, more often called "deck guns," were made by Akron Brass Manufacturing Company and were multiversal, which described their geared lateral and vertical movement. Prior to hand-turned gear controls, monitor movement was controlled simply with long handles.

Coast Guard waterfront protection included 500-gpm pump trailer units, which were towed to fires ashore. During WWII, Chrysler Corporation built thousands of these trailer pumps to bolster Civil Defense firefighting assets. Equipped with hard suction hoses for drafting from water sources, the unit used a single-stage Hale centrifugal pump powered by a Chrysler six-cylinder automobile engine.

History records that during 1942 and 1943, Coast Guard fire units responded to more than 90 percent

Called simply a Bass boat, this is one of six 36-foot fireboats built by Perry Bass Boat Works for the Coast Guard between 1943 and 1947. Like Hanley and Wheeler boats, it is piloted from a helm station at the aft deckhouse. Courtesy of US Coast Guard.

Wartime censorship omitted details of this incident off Cape Cod in April 1945. The death toll was high after this freighter collided in heavy fog with an Allied tanker. The Coast Guard vessel-turned-fireboat is of the size and configuration of a medium-endurance cutter, formerly of the navy diver class. Courtesy of US Coast Guard.

of all waterfront fires, more than one-fourth of which were discovered by USCG security patrols. In April 1942 USCG Lieutenant Arthur F. Pfister took leave from the New York City Fire Department where he was chief of the 25th Battalion (Marine Division). His new calling with the Coast Guard had him helping to organize a fleet of fireboats to assist the city's fleet in protecting nearly 600 square miles of waterfront. New York's Coast Guard fleet, which numbered sixteen red-painted vessels by the end of 1942, was under the supervision of the city fire chief, who maintained excellent rapport with the captain of the port. The USCG fleet had responded to nearly 100 fires since it received its first fireboat in July, with not a single pier or waterfront building lost. Firefighting vessels were incorporated into the Harbor Patrol Fleet and ranged in size from 40 feet to 83 feet long, with Hanley fireboats slated for arrival during 1943. Cooperation between the captain of the port and New York's fire chief resulted in an intensive ten-week course attended by all Coast Guard firefighters.

Within two months after the Coast Guard was assigned the port security mission in February 1942, a similar arrangement was established in Boston. Coast Guard captain of the port Captain Frank M. Meals and Boston fire chief Samuel J. Pope collaborated to maximize fire protection of busy Boston Harbor. Their level of cooperation not only allowed Coast

Pumps and hoses made for a crowded deck of this merchant stemwinder converted to Coast Guard fireboat. Its 4,000 gpm provided by eight pumps gave ample knockdown power for major fires in Baltimore Harbor. Each 30-foot, non-collapsible suction hose is tipped with a debris filter basket. The CGC prefix identified vessels over 83 feet long. Courtesy of US Coast Guard.

Suction hoses draft from Boston Harbor to supply the six 500-gpm pumps of fireboat CGC-905-F, a fishing trawler converted to a fireboat during WWII. A large gasoline tank on the afterdeck supplied the pumps. Courtesy of US Coast Guard.

Guardsmen to draw from the fire department's experience, some of the seven USCG fireboats and crew assigned to Boston were quartered at Boston Fire Department fireboat stations. Chief Pope also made arrangements for Coast Guard fireboat stations to be tapped into Boston's fire alarm system, with the result that at least one Coast Guard fireboat responded to every alarm at Boston's waterfront. Jeep patrols of the Coast Guard Fire Division, equipped with pump trailers, hose and chemical apparatus, often responded as well.

The seven USCG fireboats assigned to Boston were among nineteen assigned to the Northeastern 1st Naval District. The remaining dozen was dispersed among the ports of Providence and Newport, Rhode Island; Portland, Maine; Rockland and Portsmouth, New Hampshire; and Gloucester, Fall River, and New Bedford, Massachusetts. Some were converted fishing trawlers, and all were equipped with two-way radio.

It wasn't until WWII, when the Coast Guard came under control of the US Navy, that a Coast Guard service-wide vessel designation and numbering system was established, although not necessarily chronological. During 1943, Coast Guard authorities came up with a new designation system for vessels less than 100 feet long, with each vessel assigned a five-digit number, the first two numbers of which indicated the vessel's length. The third digit was the design source, with the number three or larger indicating USCG design, while those designed elsewhere were assigned numbers zero, one, or two. The last two digits identified the individual vessel within the series. These numbers were displayed on both sides of the bow and on the stern.

With patriotism sweeping the nation during WWII, in true volunteer and maritime tradition, many of

Fourth Naval District Philadelphia-based Coast Guard crewmen on October 12, 1944, give the first demonstration of a floating monitor designed for under-pier fires. Propelled by the thrust from a fire hose, the device was controlled by steering lines. The 70-foot fireboat with six 500-gpm pumps on deck was converted from the civilian vessel *Queen of Palm Beach*, which carried registration no. 241243. Courtesy of US Coast Guard.

Although owned by the City of Seattle, the powerful fireboat *Duwamish* is seen here in 1942 manned by Coast Guard 13th Naval District personnel due to manpower shortages within the fire department. Additional protection of Seattle's war-vital waterfront included thirteen types of USCG vessels. Coast Guard fire barges were also stationed at seventeen Washington ports, and at Portland, Oregon. Courtesy of US Coast Guard.

the 100,000 auxiliary and reserve US Coast Guard personnel manned more than 14,000 large craft of the Coast Guard, army, navy, and civilian fire departments. Their service not only ensured the safety of waterways in America and released Coast Guardsmen for duty overseas, they were part of the largest firefighting force ever organized, operating the largest fireboat fleet in history.

Tugboats were a mainstay of the US Navy and were well equipped for firefighting. Most famous among the navy tugs is the USS *Hoga (YT-146)*, whose crew distinguished itself during the attack on Pearl Harbor. Launched on May 22, 1941, at New York's Consolidated Shipbuilding Corporation, the 100-foot *Hoga* was assigned to the 14th Naval District at Pearl Harbor. She was underway within the first ten minutes of the December 7 attack with her crew heading for the burning ships along Battleship Row. *Hoga's* crew had all the work they could handle, heroically pulling burning ships and fighting fires for seventy-two continuous hours. *Hoga's* commendable actions that fateful week foretold her destiny—beginning in 1948 she embarked on a forty-year career as a fireboat for Oakland, California. In 1989, she was declared a National Historic Landmark—not a bad record for a tug.

When the navy re-designated its watercraft in May 1944, its tugs were classified Harbor Tug, Big (YTB), rated at more than 800 horsepower; Harbor Tug, Medium (YTM), rated between 400 and 800 horsepower; and Harbor Tug, Little (YTL), less than 400 horsepower. Standard on most tugs were diesel-driven fire pumps that delivered between 2,000 and 4,000 gpm. The number of monitors on navy tugs varied between two and five, dependent on the boat's machinery and configuration. Also standard were two 750-gallon foam tanks with the eductors fed by a 25 hp motor-driven liquid foam pump.

Often overlooked in historical annals is the US Army's inventory of watercraft, which peaked during WWII at nearly 128,000 vessels. The army was not as rigid as the navy in classifying vessels, grouping them according to size and function. A total of 718 tugboats were built for the US Army during the war by forty-six boat manufacturers. Similar to navy tugs in firefighting capability, they were constructed of steel or wood, and designated Small Tugs (ST) and Large Tugs (LT), with the latter identifying vessels longer than 90 feet. Vessels less than 50 feet long were called J Boats and categorized according to function: ambulance, rescue, patrol, and firefighting. In 1943, Higgins Industries of New Orleans, as part of its voluminous wartime boat production, built one dozen 27-foot fireboats for the army. Built with

One of the many fire prevention tasks assigned Coast Guard fireboat crews was breaking up oil slicks and flushing oil-coated pilings. These crewmen of the 50-foot fire barge CGB-50313-F use monitors to break up an oil slick in Seattle Harbor in August 1943. The barge, along with a twin (CGB-50312-F), was built by Marisen Shipyards Corp. in 1942. Courtesy of US Coast Guard.

wooden hulls and powered by gasoline engines, they were labeled Model 322 and numbered J-729 through J-740. Included among 170 wooden, diesel-powered 65-foot T-Boats built during the war were Fire Patrol Boats. Generally classified as coastal freighters, T-Boats usually were built in pairs in keeping with the wartime practice of giving work to manufacturers across the country, although all were built according to standard "Design 259" issued by the Army Quartermaster Corps. Two of these boats are known to have been built as full-fledged fireboats, wearing numbers F765 and F766. Although most T-boats fell into disuse after the war, some, including the two fireboats, remained in army service.

Built from 1943 through 1946, all eleven 60-foot USCG fire barges were assigned to the West Coast. Two detachable 286 bhp Chrysler diesel engines proved insufficient for the 43-ton vessels and they were unstable in current. The wood-hull barges had eight P-500 pumps that supplied three fixed monitors and multiple hand-operated nozzles. Courtesy of US Coast Guard.

Post-WWII

Most commandeered vessels were returned to their original owners at war's end; however, vessels that had undergone extensive alterations for firefighting were sold or scrapped. One of the privately owned fishing vessels pressed into wartime service as a Coast Guard fireboat (CG-64022-F) became war surplus and went on to serve eighteen years as a fireboat for the city of Cleveland; Boston acquired a pair of 97-foot minesweepers, which it converted to fireboats; Newark acquired a pair of Coast Guard Wheeler Boats, which served as the city's fireboats; the Mare Island Fire Department obtained a Wheeler Boat, which later went to the Eureka, California, Fire Department, and Mare Island's navy tug USS *Dekaury* became Stockton, California's fireboat. And so it went with other fire departments from coast to coast that took advantage of post-war austerity, acquiring military surplus

The US Navy tug *Cahto* (YTB-215) teams with FDNY fireboat *William J. Gaynor* and other firefighting vessels to battle fire at the Municipal Ferry Terminal Building at St. George, Staten Island, on June 25,1946. The 110-foot *Cahto*, with a crew of twelve, was commanded by Ensign James E. Hair, one of the Golden Thirteen, the navy's first black commissioned officers. The diesel-powered tug was built by W. A. Robinson of Ipswich, Massachusetts. Courtesy of US Navy.

vehicles, fireboats, and equipment at far less cost than new purchases. In some cases, equipment was donated, requiring nothing more than the formality of paperwork to complete the transfer.

Despite widespread military downsizing, military assistance at major ports did not end. Typical of navy tugs that offered fire protection to ports was USS *Tuscola (YTB-280)* that served in waters off the New England coast from 1946 to 1963. Assigned to the 6th Naval District at Boston, *Tuscola's* base of operations was the Portsmouth Naval Shipyard Fire Department, New Hampshire. For nearly a decade following the armistice, army T-Boats remained available to assist municipal fire departments. Their decommissioning began in 1953 and was completed by 1965.

For fire protection at naval bases, the navy during the 1960s ordered a small number of 60-foot fireboats powered by two 600 hp General Motors diesels. The steel fireboat had a beam of 16 feet and drafted 4 feet. Two monitors were mounted on the 4,000-gpm boats, and they carried 1,000 gallons of foam; normal crew complement was seven.

Built for heavy work and equipped with fire pumps, military tugs were easily adapted as fireboats. The main loop supplying *Oneyana's* (YTB-262) five monitors is visible beneath the upper deck. Forward monitors of tugs often were pointed rearward to prevent damage from upward bow curves of ships being pushed. The *Oneyana* was one of four YTBs built by Martinac Shipbuilding of Tacoma. Courtesy of US Navy.

In the months following the war, the US Coast Guard's firefighting role diminished, with municipal fire departments assuming responsibility using their own fireboats. Scores of Coast Guard fireboats were disposed of or became surplus to end up on fire department inventories. In 1944 the Coast Guard acquired from the army a small number of amphibious DUKW vehicles for rescue and flood operations on the East and West Coasts. Measuring 38 feet long, the DUKW's six-cylinder gasoline engine allowed a land speed of 55 mph and 8 mph on water. In 1948, Coast Guard Yard, Curtis Bay, Maryland, built additional

The *Oneyana* prepares to battle a dock fire at Pearl Harbor during the late 1940s, while in the foreground a Chrysler 500-gpm pump trailer supplies a hose line. *Oneyana* served the Hawaiian Islands for seventeen years before being converted to a floating fishing lodge in Alaska named *Sea Ranger*. Courtesy of US Navy.

An LCM-6 (Landing Craft, Mechanized) undergoes conversion to a fireboat at the 8001st Sagami Depot at Yokohama, Japan, in 1950. Sagami was a major supply depot during the Korean war. A pair of 165 hp diesel engines drove the twin screws of the 56-foot Mike Boat. Courtesy of US Army.

versions, bringing the total to twenty-eight DUKWs. The Coast Guard version was all-aluminum and featured a tow bitt and tow rail. Despite high maintenance costs and saltwater deterioration that spelled the end of their Coast Guard service, as military surplus, DUKWs became popular with fire departments that outfitted them as fire-rescue boats.

The only other amphibious craft operated by the Coast Guard was the larger LARC-V (Lighter Amphibious Resupply Cargo, 5 ton), built for the army during the early 1960s. The Coast Guard acquired twenty-seven LARCs, which it used mainly for rescue along the Atlantic coast. Although operating and maintaining the LARC was cost prohibitive for fire departments, the LARC became the basis for one of the most unique fireboat designs in fire service history.

During the late 1950s, Consolidated Diesel Electric Corporation of Connecticut developed the LARC-V for the US Army to transport troops and supplies from offshore ships to inland areas. The LARC entered military service during the early 1960s, however, compared to the DUKW, saw limited use, mainly in Vietnam; less than 1,000 were built. Fort Story, Virginia, became home to the LARCs, while some examples went to the US Navy, which continues to operate them on a limited basis.

Despite its all-aluminum construction—having a 3/16-inch-thick hull—the LARC weighed in at a hefty 10 tons and had a cargo capacity of 5 tons. The single-screw, four-wheel, self-propelled LARC measured 35 feet long and 10 feet wide. Power was supplied by an eight-cylinder Cummins V-8 diesel engine rated at 295 bhp at 26,000 rpm. The engine was in the stern over the propeller, with drive directed forward to a transfer case that enabled two- or four-wheel drive. The LARC's huge, low-pressure tires not only permitted travel on soft surfaces and steep grades, they served as the only suspension system and gave the bulky amphibian superior stability in the water. The LARC traveled at 30 mph on land and 10 mph in the water.

Unquestionably, the most interesting, if not unusual, conversion of military equipment to fireboat involved a small number of LARC-Vs. After a firsthand look at the LARC-V operated by Coral Gables, Miami Fire Department Division Chief of Support Services Louis DeChime envisioned a specialized, multi-purpose, amphibious vehicle for the fire service. Since the DUKW's replacement in the military was the LARC-V, DeChime followed suit, teaming with Ladder Towers, Inc., (LTI) of Ephrata, Pennsylvania, to produce the amphibian.

Prior to LTI's LARC-fireboat conversions—which went to Milwaukee and Tampa—a number of agencies,

Few army T-Boats are known to have been purpose-built as fireboats. Flanked by other army boats, F765 shows off its colorful scheme during the 1950s. The Quartermaster ensign flies from the yardarm. The silver-painted gasoline tank atop the deckhouse was typical of fire pump-equipped tugs of the period. Courtesy of US Army.

both government and commercial, became involved with a mammoth project to develop a uniquely effective and powerful LARC-fireboat conversion. Although the LARC fireboat project involved numerous players during its long life, DeChime points out that it was "his baby" during his thirteen years of commitment.

The story begins with the US Coast Guard, which, despite having been out of the fireboat business for more than thirty years, continued research to improve marine fire safety. During the 1970s, Coast Guard officials teamed with the Technology Utilization Office of the National Aeronautics and Space Administration (NASA) to examine specific methods of marine firefighting. Such commitments were well within the parameters of NASA's engineering of safety systems. In the wake of disastrous fires aboard oil tankers, in 1975 a program was begun to investigate ways to fight these massive fires.

In 1978, NASA's Marshall Space Flight Center at Huntsville, Alabama, began working jointly with the Coast Guard to develop a firefighting module to combat shipboard and harbor fires. Overseeing NASA's involvement was its director of Technology Utilization, Ismail Akbay, who, years earlier, served as a lead engineer on Apollo/Saturn V rockets. Aviation Power Supply of Burbank, and later the Aviall Company, was selected to build the module under NASA license. In March 1982, Kinetics Technology International (KTI) of Santa Ana, California, purchased the rights to the unit, which was named Firefly. Engineers at NASA's Marshall facility drew from their extensive experience from space projects involving high-capacity pumps, lightweight materials, and compactness to design a unit with everything needed for fighting fires, including engine, pump, hose, and fittings. It was to be transportable by helicopter or lifted by crane to be placed aboard a vehicle or vessel, or at dockside. The Coast Guard tested the unit aboard a patrol boat for rapid-installation auxiliary firefighting. The navy tested it loaded onto a pickup truck, and placed the unit by helicopter aboard a tugboat under way.

This 64-foot all-wood harbor tug was one of fifteen built during 1943 and 1944 for duty at New York and Philadelphia. Lacking an integral fire pump, they were outfitted with a P-500 pump atop the deckhouse. A Murphy 160 hp diesel engine gave the boat a top speed of 13 mph. Number 64305, on the Delaware River in 1955, was one of six of the type built by Patchogue Basin, New York. Courtesy of US Coast Guard.

Firefly I's unique mix of components gave designers the lightweight, compact, yet powerful unit they were looking for. At the heart of the unit was an Ingersoll-Rand 2,500-gpm, two-stage pump. Especially noteworthy was the engine, which had earned its reputation powering the army's highly successful OH-6A Cayuse Light Observation Helicopter (LOH). Built by Allison of Detroit Diesel, the Model 250KB gas turbine engine was rated at 250 horsepower, with a maximum output of 400 horsepower. Battle tested in the OH-6A during the Vietnam War, the engine proved many times over that it could easily exceed 100 percent operating power. The powerplant's performance record, plus its mere weight of 158 pounds, made it the ideal choice for the Firefly module. The use of aluminum, titanium, and stainless steel held the module's weight to 2,500 pounds, 1,000 pounds of which was fuel for four hours of operation.

The Coast Guard decided not to purchase the Firefly module, but later developed an experimental version that not only pumped 5,000 gpm for firefighting, but could be used to de-water ships and establish a water supply on land. The US Navy purchased two Firefly modules, one of which was trailer-mounted for use at the Philadelphia Navy Yard. For commercial applications (for which there seemed greater interest), the module was seen trailer-mounted, skid-loaded onto pickup trucks, and tested aboard a St. Louis fireboat. Dow Chemical's Texas Division at Freeport mounted a Firefly aboard its fireboat. An improved version, which was fully developed by 1981 and named Firefly II, was capable of pumping 3,000 gpm from a hydrant system or from bodies of water.

Louis DeChime took a keen interest in the concept, which led to his mating the Firefly with a LARC for Miami. Combining military amphibious technology with high-performance pumping technology from NASA created a new firefighting weapon that functioned as fireboat and conventional land pumper. Watching closely with more than a passing interest were officials of the fire departments of Buffalo, New York City, and Jersey City, New Jersey, the details of which are covered in sections devoted to those cities.

In 1945, the 86-foot tug ST-850 was delivered to the army by Tampa Marine, which built eighteen STs (Small Tugs) for the US Army. At Inchon, Korea, in 1954, the tug sports a fire monitor tower atop the pilothouse for maximum stream reach. Courtesy of US Army.

The 65-foot US Army fireboat F766 prepares to operate its two monitors off Fort Eustis, Virginia, in 1955. Courtesy of US Army.

The T 45 was one of four built for the army by Lester & Franck Shipbuilders of Seattle. Fully functional as a fireboat and aircraft rescue boat, it is seen here on Lake Washington near Boeing's Renton Plant in 1952. Close by is a 1947 Aeronca S11BC of Washington's Lake Union Air Service. Courtesy of US Army.

Named in honor of WWII hero US Army Sergeant Robert D. Farmer, who was killed in combat, T 205 had been built by Minneford Boatyard at City Island, New York. The *Farmer* served Whittier, Alaska, which had been established by the army during the war. The deep-water military Port of Whittier remained active until 1960. Here, the *Sgt. Robert D. Farmer* works at the blaze that destroyed the entire Alaska Railroad Dock on June 17, 1952. The yard master jumped 35 feet from the burning dock to safety on *Farmer's* deck. Courtesy of US Army.

The US Coast Guard relied heavily on its helicopters to support its vessels and municipal fireboats, often arriving ahead of them at emergencies. This Sikorsky HH-52A Seaguard approaches a shrimp boat afire off Miami's Dinner Key in August 1964. Two shrimpers who were burned when the boat's gasoline tank exploded were pulled from the water by Coast Guardsmen. Courtesy of US Coast Guard.

Scale models in the Louis DeChime collection show the variations in engine and plumbing distribution configuration of the three LARC fireboat conversions. From front to rear are the amphibious fireboats of Tampa, Miami, and Milwaukee. Courtesy of Delia DeChime.

Besides Miami, Milwaukee, and Tampa, other fire departments known to have operated LARCs are those of Virginia Beach, Virginia; Reading, Pennsylvania; Nantucket, Massachusetts; Sodus Point, New York; and Port Clinton, Ohio. The Toledo, Ohio Fire Division in 1973 acquired a LARC in the aftermath of the city's worse flood, during which homes burned but couldn't be reached by land apparatus. The fire department shop mated a modified 750-gpm Hale pump with a diesel engine to supply a monitor. Although successful, the LARC fireboat served for only two years. The Norfolk Naval Base Fire Department obtained a LARC and equipped it with a portable fire pump and rescue equipment. A LARC with Florida's Coral Gables Fire Department, which came from navy surplus at Key West, would play an important part in DeChime's fireboat venture.

Despite having dissolved its fireboat fleet after WWII ended, the US Coast Guard maintained some degree of firefighting capability in all of its vessels. Among a wide variety of Coast Guard fire pump-equipped vessels ranging from 30 to 140 feet long, the 32-foot Port and Waterways Boat (PWB) proved to be the most capable for its size when it came to fighting fires. In response to the Port and Waterways Act of 1972, fifty PWBs were built in 1976 by the Willard Boat Company of Santa Ana, California. Numbered 32301 through 32350, the boats featured a 500-gpm pump driven by its own Detroit diesel engine, while twin 406 bhp Caterpillar diesels drove twin screws. A Feecon 500-gpm monitor on the roof of the pilothouse was remotely controlled. With hull and superstructure constructed of fiberglass, the PWB had a top speed of 23 mph. PWBs served at ports across the country until the late 1990s, when they were made available as surplus; fire departments were then quick to acquire the top performers as fireboats.

True to its age-old commitment to service, when America was attacked on 9/11, more than one dozen US Coast Guard vessels, from various utility and rigid inflatable boats to 65-foot harbor tugs and 140-foot Bay Class ships, were counted among the long list of vessels that responded.

Downsizing of the army's massive fleet of vessels over the years, which included T-Boats and fireboats, left but a handful of boats equipped for firefighting. Noteworthy among them is the *Valley Forge (LT-2096),* which ended its career in 2011 as the army's oldest tug. Weighing 141 gross tons and stretching 108 feet, *Valley Forge* spent its entire career as the fireboat at Military Ocean Terminal Sunny Point (MOTSU), Southport, North Carolina, the army's largest ammunition port and its primary East Coast deep-water port. The *Valley Forge* was built as a dual-purpose tug and fireboat during 1954 by Matton Shipyard at Cohoes, New York. In early 1955, *Valley Forge* left the Matton yard as Hull No. 308, powered by two Detroit diesel engines and two Caterpillar

Weighing 141 gross tons, the dual-purpose tug and fireboat *Valley Forge* (LT-2096) was built by Matton Shipyard, Cohoes, New York, as hull number 308 and delivered to the US Army in July 1954. It is seen here in 1999 at Hythe, England, following rebuild. When *Valley Forge* was retired in 2011, it was the oldest tug in army service. Courtesy of Dr. Allan Ryszka-Onions.

engines, giving her a top speed of 15 mph. Two monitors easily handled 4,000 gpm of her total 7,500-gpm output, along with 1,250 gpm through a foam monitor, fed from a supply of 1,000 gallons of AFFF (aqueous film forming foam). Rebuild at Hythe, England, in 1999 included side-thrust steering and an electronics suite. Her crew was convinced that the army motto, "Army Strong," was coined for *Valley Forge*. But *Valley Forge's* fate was sealed when army officials decided that the thinness of her hull did not meet regulations, plus a replacement fireboat had long been in the works. When she went to government liquidation, *LT 2096* was in pristine condition and her engines were like new. *Valley Forge* sold at auction in 2012 for considerably more than its $150 starting bid—her pumps alone were valued at more than $1 million—and went on to a new career herding scrapped vessels to coordinates where they were sunk to form artificial reefs off the Atlantic coast.

Replacing the *Valley Forge* at Sunny Point was a purpose-built fireboat designed specifically by army engineers. The 76-foot, all-aluminum vessel does not share the pumping rating or speed of fireboats its size, which was in keeping with the army's wishes. When army officials went to Gladding-Hearn Shipbuilding at Somerset, Massachusetts, they had done their research and stated that they did not want a lightweight, fast boat with pumps run off the main engine. What they got was a deep-V hull boat capable of simultaneous, constant 100 percent pumping and propulsion.

Designed jointly by the shipyard and C. Raymond Hunt Associates, the vessel's firefighting system consists of two pumps, each powered by a Caterpillar 503 bhp diesel engine. Their separation from twin propulsion engines ensures maximum firefighting capability and maximum power. Power is derived from Caterpillar diesels, each producing 1,600 bhp to turn a pair of oversize Hamilton waterjets for stable maneuvering. Peak output is 7,500 gpm to

Using environmentally safe colored vegetable dye, *Valley Forge* salutes during the dedication of Military Ocean Terminal Sunny Point's (MOTSU) Center Wharf at Southport, North Carolina, in May 2010. MOTSU is the largest military terminal in the world, with the *Valley Forge* operated by its Directorate of Emergency Services. Courtesy of US Army.

With ample fendering protecting its hull, the navy tug *Eufaula* (YTB 800) escorts the carrier *America* in June 1982. It was a Natick Class District Harbor Tug launched by Southern Shipbuilding Corp., Slidell, Louisiana, on February 22, 1969, for the 5th Naval District at Norfolk. *Eufaula* measured 109 feet with a 31-foot beam, and displaced 356 tons. Courtesy of US Navy.

The *Omeed Taher* is one of two fireboats chartered to the Kuwait Ports Authority under the US Army. It is alongside its partner fireboat, *Shareif Verve*, displaying its firepower in the Port of Shuaiba in September 2009. Courtesy of US Army.

supply five monitors and eight hydrants. Monitors on the fore and aft decks are manually operated, while a 5,000-gpm monitor atop the pilothouse is remotely controlled. In addition, two 200-gallon tanks of foam concentrate supply a 1,200-gpm foam system. The state-of-the-art vessel was delivered to Sunny Point's federal fire department in 2011.

The army's fireboat at MOTSU is an exception to the small number of pure fireboats operated by the US armed forces in the twenty-first century. The priority given protection of military shipping sites is also evident in a contract awarded in May 2014 to Technology Associates,of New Orleans. The contract called for a high speed, 75-foot, 10,000-gpm army fireboat to protect the Military Ocean Terminal at Concord, California (MOTCO). The navy has significantly trimmed its tugboat inventory, with much of its tug work done under contract with commercial tug firms. Even military fireboats that protect ports of American interests abroad usually fly that nation's flag, yet operate under US charter. Throughout history, the Coast Guard has consistently maintained the largest fleet of US military vessels with firefighting capability, and expects to continue that service.

CHAPTER 2: GREAT LAKES

Although many of the details of disastrous fires that ravaged the Great Lakes region have been relegated to history, the toll they took in lives and property lost is not to be underestimated. As far back as the mid-1800s, fireboats played an important role in travel on the lakes. Ships that traversed the lakes helped build the great cities of Chicago, Duluth, Milwaukee, Detroit, Cleveland, and Buffalo, among others. They brought droves of immigrants from Europe to settle in the region or to make their way westward. Besides passengers, these vessels hauled coal, iron, and agricultural products. Often their cargo was combustible, having been harvested from Midwestern forests.

Beside the infamous, unpredictable storms that wreaked havoc with heavily loaded vessels on the Great Lakes, fires and boiler explosions, which often occurred at or near ports, accounted for the heaviest losses during the 1800s. The ports themselves, which were long void of safety regulations, held the potential for fire during their growth. Too often, that potential became evident in acres of smoldering ruin, highlighting the need for fireboats.

The largest loss of life on the Great Lakes occurred on 24 July 1915, when the top-heavy passenger steamer *SS Eastland* rolled onto its side while docked on a Chicago river. Although there was no fire, Chicago firefighters and fireboat crews joined throngs of rescuers in the grim task of retrieving 848 bodies.

First to place a fireboat in service on the Great Lakes was Captain Richard Armstrong, an experienced Great Lakes sailor. Armstrong moved to Bay City, Michigan, in 1869 to tap the wealth of the booming lumber industry by establishing a boating business to keep pace with the city's growth. Bay City, at the base of Saginaw Bay on Lake Huron, quickly developed a reputation of being a shipbuilding community, beginning with several small shipyards, followed by three major shipbuilders, James Davidson, Frank W. Wheeler, and Harry J. Defoe, all of whom would build fireboats.

In 1881, Armstrong placed in service the first fire tug on the Great Lakes, and within four years added four fireboats to protect waterfront lumber mills. Armstrong ran his fleet under the name Saginaw River Fire Boats, most of which were built by Davidson and Wheeler. Throughout the 1880s, Armstrong's fireboats were assigned specific zones in which sawmill owners contracted for fire protection; contracts included the line: "No assistance rendered to parties not contributors." Popular among his fireboats was the all-wood *Geyser* (not to be confused with its namesake of the Chicago Fire Department), built in 1886 as Davidson's hull number 27. Armstrong's other fire tugs were named *C. M. Farrar, Chas. Lee, Handy Boy,* and *David Sutton.* He later added a deep water passenger service and renamed the business Bay Line and Fire Boat Company, which was more popularly called Boy Line since many of the boats' names included the word boy, such as *News Boy* and *Handy Boy.* Railroad expansion, a decline in waterfront lumber mills, and the growth of fire departments during the 1880s caught up with Armstrong, causing his enterprise to dwindle.

The *Geyser* was built in 1886 by James Davidson for Captain Richard Armstrong's private fireboat service on the Saginaw River. Weighing 29 gross tons, it was assigned US registry number 86036. Visible on its bulwark is the title BL & FB Co., for Bay Line and Fire Boat Company. Author's Collection.

A similar endeavor existed, although on a smaller scale, at the twin cities of Menominee, Wisconsin, and Marinette, Michigan, with the wooden fire tug *Menominee River.* Construction of the 78-foot vessel in 1879 is credited to both John Gregory and A. J. Johnson of Green Bay. Around the turn of the century, following ownership by Kirby Carpenter Co. of Marquette, Michigan, she went to Menominee River Fire Tug Co. of Marinette.

Chicago

No US city seems to identify historically with fire more than Chicago. As David Cowan states in his book *Great Chicago Fires*:

> Certainly no other city cites a great conflagration as the cornerstone of its will and identity. Yet the Great Chicago Fire was not the only infamous blaze the city would see. Rather, as Chicago changed from agrarian outpost to industrial giant, it would be visited time and time again by some of the worst infernos in American history.

Preceding the fire that started on October 8, 1871, in the barn of Mrs. O'Leary's fabled cow, the fire chief's request for a fireboat had been denied. The tugboats that city fathers deemed more cost effective to meet Chicago's firefighting needs were helpless in checking the fire's spread to the industrial district along the Chicago River. Although it's doubtful a fireboat would have stopped the blaze, it likely would have reduced its spread and provided a water supply, thereby reducing the number of people killed and left homeless. And so, the Windy City, like other major port cities, would learn hard lessons, yet six years would pass before waterfront fire protection appeared. In 1877 the city began using three fire tugs on the Chicago River, two of which were stationed to cover the lumber district. A later addition was the former 77-foot lumber tug Alpha, which was organized as Engine 37 of the 8th Battalion on September 5, 1885, at Allen's Slip. Engine 37 was Chicago's first fireboat company, which would be joined by Engines 41, 58, 71, and 92. The need for additional fireboats was obvious, prompting Chicago Fire Department Fire Marshall and fireboat advocate Denis J. Swenie to conduct a fact-finding tour at New York City and other East Coast fire departments. On his return, work on the *Geyser* began at Miller Bros. Dry Dock Co. in Chicago. It emerged in July 1886 at 143 gross tons, measuring 98 feet with a 23-foot beam and drafting 10 feet. A pair of Clapp & Jones piston pumps powered by a Scotch boiler produced 3,810 gpm.

The *Geyser* replaced the *Alpha* on November 29, 1886, and was stationed near Throop Street Bridge. Relocating the fireboat more than a decade later to the newly built Ogden Canal that formed Goose Island marked the first time the Chicago Fire Department was able to fight fires from the river. The *Geyser* would respond to all fire boxes within two blocks inland of the river. Impressed by the *Alpha's* performance at the Chicago Lumber Co. fire in May 1885—the city's largest blaze since the 1871 Great Chicago Fire—lumber merchants chartered the fire tug and provided fuel, while firefighter crew salaries were paid by the city. Built in 1881 as a G-1 style tug, the *Alpha,* when rebuilt, produced 3,000 gpm with pumps installed by Chicago's Chas. F. Elmes Engineering Works.

The Chicago Fire Department was quick to show the city that it got its money's worth in the *Geyser.* During the winter following her arrival, the fireboat, with its bow reinforced for ice-breaking, kept open the Chicago River, convincing officials that a major flood had been averted. The *Geyser* and her crews also received high marks for their work at major blazes, which cleared the way for the purchase of additional fireboats to keep pace with Chicago's industrial growth. Next to arrive was the *Chicago,* an 81-foot former lumber tug rated at 2,260 gpm. It was built in 1886 and mounted a single monitor. Four years later the *Yosemite* was launched at Miller Bros. yard on the Chicago River. The 98-foot vessel weighed 141 gross tons and could pump at a rate of 5,000 gpm. As often happens in historical documentation, details of *Yosemite's* identity are contradictory. While it has been noted that its name was changed to *Protector* in 1903, historical records of Great Lakes vessels held at Bowling Green State University show that they were, in fact, two separate

The Chicago fireboat experience began in 1877. The *Yosemite* served as the city's seventh vessel for protection of the growing waterfront. The 500-gpm fireboat began service in 1890 and served sixteen years before retirement. Author's Collection.

fireboats. The *Yosemite,* which was assigned US registration number 200604, was officially struck from the fire department's inventory on March 2, 1906. The fireboat *Protector,* which was assigned registration number 203110, was launched two months later at Buffalo, New York.

The World's Fair Columbian Exposition held in Chicago during 1893 not only celebrated the 400th anniversary of the discovery of the New World by Europeans, the six-month event showed the world that Chicago had risen from the ashes of the Great Chicago Fire of 1871. It seemed fitting, then, that a special fireboat was built for the global event, which bordered the lakeside and included three miles of canals and lagoons. Designed by the exposition firm, the fireboat, named *Fire Queen,* began construction in 1892. To negotiate the fair's quiet waters, she was a shallow-draft, Doyle G-4–type vessel. Its pump rating was 2,100 gpm, and although no monitors were mounted, the boat had four 3½-inch hose connections. The *Fire Queen* was berthed near the Manufacturers and Liberal Arts Building, then the largest building in the world. When demolition of the 630-acre site began in 1894, the fireboat was turned over to the city and stationed at Goose Island, along with Engine Company 63. The *Fire Queen* was called Engine 71, with the designation later passed to the *Geyser.*

The arrival of Chicago's first all-steel fireboat brought the total to five. Built in 1898 as Chicago Shipbuilding Company's Hull Number 34 and named *Illinois,* the 107-foot vessel boasted an output of 10,000 gpm thanks to a pair of duplex reciprocating pumps fed by two Scotch double boilers. With the ability to put 11 master streams on a fire, only *The New Yorker* could out-pump her. With such power, and weighing the heaviest at 287 gross tons, with a 24-foot beam and 12-foot draft, the *Illinois* easily reigned as flagship of the fleet. Her bow deviated from traditional style in being sloped to allow the boat to ride up onto ice and then break through.

A number of changes in Chicago's fireboat fleet occurred at the beginning of the twentieth century. On January 1, 1903, the *Geyser* was fittingly renamed *Denis J. Swenie.* Retirement of the *Fire Queen* in 1905, followed by the *Yosemite* the following year, generated plans for a new fireboat. The result was a 95-foot wooden fireboat named *Protector,* built by Empire Shipbuilding Co. of Buffalo, New York. Before embarking on an eighteen-year career as a Chicago fireboat, on January 17, 1907, it was renamed *Michael W. Conway* after a battalion chief during the 1800s, who was one of the first firefighters to put water on the Great Chicago Fire.

Formerly the fireboat *Protector,* the *Michael W. Conway* was a wooden 95-footer that served eighteen years as one of Chicago's early fireboats. Author's Collection.

In August 1908, the mettle of Chicago's three ablest fireboats and crews would be tested against a conflagration that could have rivaled the Great Chicago Fire. The site of one of the nation's largest assemblage of grain elevators, yet lacking an adequate water supply and access for fire apparatus, Chicago's south branch of the Chicago River was a recipe for disaster. A discarded cigarette caused a massive explosion in, of all things, an explosives factory, which ignited huge nearby grain elevators. Wind pushed the firestorm at the Burlington Railroad yards southward toward more towering elevators, a massive warehouse, and two oil refineries. The *Illinois* held a position to battle the inferno, while crews of the *Chicago* and *Swenie* did likewise. Collectively, their monitors totaled only four to throw 20,000 gpm against the fire, yet the three fireboats stood fast as radiant heat blistered their paint and cracked windows. Twice, flames roared over the boats and started fires on the opposite bank. The following morning, as firefighters made headway, a massive explosion caused an elevator several hundred feet tall to collapse onto the *Illinois,* severely injuring a firefighter and sinking the vessel. Chicago Fire Chief James Horan vowed improvements after the conflagration, with his first priority being to get the *Illinois* off the bottom of the Chicago River. Two hose wagons, each carrying 1,000 feet of 2½-inch hose, would be added for city-wide response. Not only was the *Illinois* raised and repaired, it remained in fire service until 1941 and, as of this account, works as a tug with its fifth commercial company. If not for the bravery and superhuman efforts of the fireboat crews, the city would have suffered a replay of 1871.

In the wake of the 1908 fire and others along Chicago's waterfront, there was little resistance to

The Chicago fireboat *Illinois* being raised from the river bottom after it was sunk by a collapsing grain elevator during a massive fire in August 1908. The *Illinois* not only remained in service as Engine 37 until 1941, it began a long career as a commercial tug. Author's Collection.

the acquisition of larger, more modern fireboats. Work began almost immediately at Wisconsin's Manitowoc Dry Dock Co. to build two identical steel-hull, 107-foot fireboats. Assigned company hull numbers 22 and 23, they were named *Joseph Medill,* after Chicago's mayor from 1871 through 1873, and *Graeme Stewart,* after the president of the Chicago Board of Education from 1853 through 1905, respectively. The *Medill* was not officially named as such until 1911, having first been named the *A. E. Ackerman.* The pair would be like no other, having innovative features that set them apart from fireboats of the period. At the launching of the *Medill,* Chief Horan said:

> These powerful floating engines are indispensable nowadays, as we all saw at the Burlington Railroad fire last August. These fireboats can do the work of twelve engines and the volume of water they provide is far more effective than many streams from nozzles of smaller gauge. We will also use these fireboats to supply engines in areas with low water pressure that are in close proximity to the river (*Chicago Sun-Times* 1909).

Both boats were powered by Curtis turbines with steam supplied by Scotch boilers to provide 9,000 gpm at 150 psi. Each boat sported two monitors—two amidships on a raised platform of the *Stewart*, and one at the *Medill*'s bow and one amidship. Incorporated into the base of monitors were nine 3½-inch valves to supply nozzles temporarily mounted along the taffrails. Variable-speed reversing motors were directly connected to twin propeller shafts, with speed and steering controlled from the pilothouse. The twin screws gave the pilot the option of using them in combination with the rudders, or total steering with screws only. Built in typical Great Lakes low-profile style, the boats could fit under Chicago's many bridges. Unique to the twins were three stabilizer spuds, a concept borrowed from work barges; two were immediately abaft the pilothouse and a third aft, all of which could be operated with steam-powered rack-and-pinion mechanisms. Tipped with spikes, spuds could be lowered 24 feet to the river bottom to anchor the vessel in place during pump operations.

The *Graeme Stewart* at the ready in 1930. It weighed 309 gross tons and carried US registration number 206149. A twin to the *Joseph Medill*, both featured steam turbine technology and twin screws for maneuverability. After being retired from the Chicago Fire Department in 1942, the *Stewart* worked for three Canadian tow companies before being scrapped in 1970, capping more than six decades of service. Author's Collection.

Although they solved the problem of committing monitor streams to counter the thrust of those used against fire, the spuds ruled out rapid maneuverability, an absolute necessity in fighting major blazes. The huge spuds made the vessels top-heavy and their mechanical operation proved troublesome, especially in freezing weather. Drawbacks aside, the Chicago Fire Department was pleased with the new fireboats, while hopeful that the older *Illinois*, *Swenie*, *Conway*, and *Chicago* would undergo repair. The *Chicago* was soon scrapped and the *Conway* was retired in 1924. The *Swenie* was towed out into Lake Michigan on December 29, 1927, and scuttled.

More than a quarter of a century passed before another fireboat keel was laid for Chicago. More firepower was needed to combat the significant number of waterfront blazes and relief was needed for aging fireboats; the *Illinois* was nearing the forty-year mark and was slated for replacement. In 1936, fire department officials went to Defoe Shipbuilding at Bay City, Michigan, with specifications for an all-steel, low-profile, powerful fireboat. The company had been formed in 1905 as Defoe Boat & Motor Works by the unlikely team of Harry J. Defoe, who was a public school principal; his brother Frederick W. Defoe, a New York lawyer; and their brother-in-law George H. Whitehouse, who was in the wholesale fish business. Their yard turned out mainly US Navy vessels but they welcomed Chicago's fireboat design. With war clouds forming, the timing could not have been better for both customer and builder; Chicago would become a major port stockpiled with war materiel, and Defoe would soon be busy with government orders.

Service tests of the *Graeme Stewart* in 1909 included pumping to supply eight rail standees. Visible at left is the splined portion of the unique spud that could be lowered to the bottom to anchor the boat while pumping. Author's Collection.

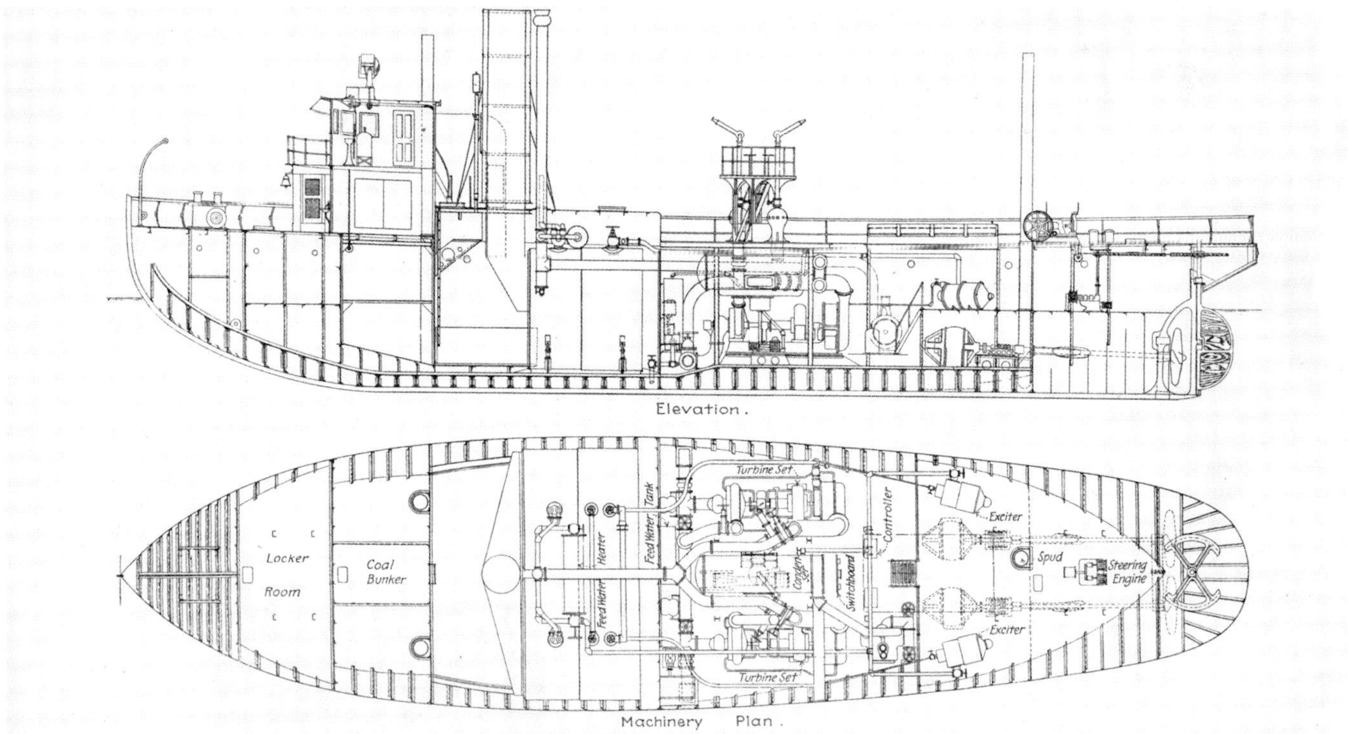

FIG. 1. PLAN AND PROFILE OF FIREBOAT "JOSEPH MEDILL."

Profile of Chicago's *Joseph Medill*, which appeared in *Engineering News* on December 3, 1908. Author's Collection.

The crew of the *Medill*, a hardy lot, took time out in January 1930 to pose alongside the vessel's pilothouse. Grim looks reflected the city's announcement that fireboat crews would be out of work due to budget cuts brought on by the Great Depression. Seen to good advantage are the spuds in their raised position. Courtesy of *Herald and Examiner*.

Acting Commissioner D. J. Carmody sounds the *Medill's* steam whistle to signal the start of Let's Go Chicago week during May 1933. Huge cowl ventilators, like those between the fireboat's spuds, eventually became a thing of the past, thanks to air conditioning and mechanical ventilation systems. The detailed emblem of the Chicago Fire Department adorns the stack. Courtesy of *Chicago Tribune*.

When completed, Defoe's hull number 156 was delivered to Chicago in April 1937 and went in service on May 1. It was named the *Fred A. Busse* after Chicago's mayor from 1907 to 1911 and designated as Engine Company 41. Weighing 157 gross tons, the *Busse* measured 90 feet, with a beam of 22 feet and an 11-foot draft. Four Dean-Hill three-stage centrifugal pumps powered by four of its five diesel engines gave her a pump rating of 10,000 gpm at 150 psi. The boat's punch was delivered through two bow monitors, one at the stern and one in a tower that could be elevated 27 feet above the water-line. The tower was immediately abaft a long deckhouse that could accommodate a crew of eight.

With the *Medill* taken out of service in 1941, and the *Stewart* in 1942, the Chicago Fire Department welcomed assistance from the US Coast Guard to augment fire protection of its waterfront during WWII. In addition to Coast Guard fireboats of the 9th Naval District, the Chicago area Sea Scouts donated their 36-foot vessel for conversion to fireboat.

Three nozzle stands and a rail standee nozzle at the stern add to the punch delivered by the *Busse's* four fixed monitors. During the 1937 dockside demonstration, the *Busse* showed off its elevating platform, which placed the monitor 35 feet above the water. The ladder moved with the platform, its base sliding inside a track atop the deckhouse. Courtesy of Wide World Photo.

The *Joseph Medill* goes to work against the Rosenbaum Bros. grain elevator fire at 102nd Street and the Calumet River on May 11, 1939. The boat's afterdeck spud is lowered to the river bottom to prevent the boat's rearward movement from nozzle reaction. Author's Collection.

The *Busse* when she was new in 1937, assigned Chicago Fire Department shop code G-006. While berthed, a connection was maintained with a fire alarm box at pier side. After retirement, the *Busse* began a new career as a cruise boat in Door County, Wisconsin. Aboard is a small museum dedicated to Chicago firefighting. Author's Collection.

With officials crammed aboard on April 2, 1942, the Chicago Fire Department's *Sea Sparke* undergoes inspection trials. To bolster the fleet following fireboat retirements, local Sea Scouts loaned the former Coast Guard motor lifeboat to the fire department. A total of 134 of the diesel-powered boats were built for the Coast Guard. The CFD shops converted the *Sea Sparke* to a fireboat by adding two pumps and three monitors. Author's Collection.

The *Sea Sparke* packed considerable punch in its 36-foot frame, as seen by this view on the Chicago River shortly after it went in service in April 1942. Author's Collection.

The Chicago Fire Department would get its second set of twin fireboats during the postwar era, when the trend of big boats was winding down. Designed by noted naval architect John G. Alden, construction of the all-steel 92-foot fireboats began in 1948 at the Christy Corporation, Sturgeon Bay, Wisconsin. The company had changed its name the previous year from Leathem D. Smith Shipbuilding, as it had been known since its beginning in 1918. In 1967, Christy and the adjacent yard, Sturgeon Bay Shipbuilding, were acquired by Manitowoc Corp., which then became Bay Shipbuilding Company. Hull number 365 would be Chicago's second fireboat bearing the name *Joseph Medill,* while its twin, hull number 366, would be christened *Victor L. Schlaeger,* in honor of the WWI veteran and Cook County recorder of deeds who died in 1949. Both boats were completed 100 days ahead of schedule, and on delivery, the *Medill* was designated as Engine 37 and the *Schlaeger* as Engine 58.

The *Joseph Medill* launches while the *Schlaeger* awaits its turn at Sturgeon Bay, Wisconsin, on November 5, 1949. The *Medill* was christened by Mrs. Michael J. Corrigan, the wife of Chicago's fire commissioner, and the *Schlaeger* was christened by Grace Louise Schlaeger, daughter of Victor L. Schlaeger. Their low profile was typical of Great Lakes fireboats. Courtesy of Door County Maritime Museum.

Chicago's second set of twin fireboats arrive at the Michigan Avenue Bridge on the Chicago River on November 15, 1949. The *Victor L. Schlaeger* was assigned CFD shop code G-004, and the *Medill*, the Windy City's second so-named, shop code G-007. Large deckhouses that incorporated pilothouses accommodated a crew of eight that included pilot, officer, two engineers, and four firefighters. Author's Collection.

In preparation for an even larger display during Fire Prevention Week 1951, the fireboat *Joseph Medill* demonstrates its pumping ability on September 29 on the Chicago River at Franklin Street. The fireboat's pumps supplied thirty-six 2½-inch hose lines. Pumping ability is standard proof of fireboat performance measured in terms of gallons-per-minute, or gpm. Author's Collection.

The two boats were powerhouses, each boasting eight 200 hp General Motors 6-cylinder diesel engines driving twin single-stage 3,000-gpm centrifugal pumps and four dual-stage 1,500-gpm pumps. Each weighed 209 gross tons and measured 92 feet, with a 24-foot beam and 7-foot, 6-inch draft. Nineteen discharge outlets and 3,600 feet of hose allowed their frequent use as pumping stations for heavy stream land vehicles and hand-lines. Six monitors, including one atop a tower that could be elevated 35 feet above the waterline, easily managed the 12,000-gpm rating, although 14,000 gpm was possible.

Reflecting true Great Lakes design, the boats were only 15 feet, 7 inches high, which left fully 15 feet of clearance under bridges without having to wait for them to be opened. Also typical of Great Lakes fireboats were reinforced bows for ice-breaking. The *Medill* and *Schlaeger* could break foot-thick ice, thanks to ¾-inch-thick bow plates added to their ⅜-inch thick

Like land-based fire apparatus, fireboats normally were maintained in pristine condition above and below decks. Fireboat engineers were masters of their trade, taking pride in their boat and their ability to provide the water necessary to fight major blazes. Engineers Frank D'Arcy (left) and Jerry Flynn work at the controls of the *Victor L. Schlaeger's* eight diesel-turbine powerplants. Author's Collection.

hulls—"bubblers" surrounding the boats kept the water from freezing so they could get a running start in the dead of winter. The boats' top speed was 16 mph.

The *Medill* was placed in service on December 2, 1949, and the *Schlaeger* on January 16, 1950. Long deckhouses provided live-aboard quarters for a crew of eight, which was later reduced to five. Locations of the big boats were based on the amount of traffic on the city's waterways. *Engine 37 Medill* operated from Navy Pier during summer months when shipping was at its peak and moved to a downtown berth in winter. Engine 41 *Busse* was stationed inland on the Chicago River, shifting closer to downtown in summer. Engine 58 *Schlaeger* worked at the Calumet River Port, where shipping thrived in summer.

Satisfied that their large, powerful fireboats provided ample coverage of Chicago's waterfront and rivers, Chicago Fire Department officials at the end of the 1950s turned their attention toward enhancing the fleet with smaller, more maneuverable boats. The department's first jetboat began service in 1960, berthed with the fireboat *Medill* at Navy Pier. This boat crew picks up a diver in a maneuver similar to that practiced by navy divers. Courtesy of Curt Elie Collection.

A proponent of the Civil Defense Fire & Rescue Service, Chicago fire commissioner Robert J. Quinn, aboard the fireboat *Busse*, watches a demonstration of a 16 foot boat driven by Robert Rodi on September 13, 1960. The fast, pump-equipped boat was donated by Rodi Boat Co. to the fire & rescue unit. Author's Collection.

The appointment of Robert J. Quinn to Chicago Fire Commissioner in 1957 brought with it a whirlwind of change. The colorful Quinn's innovations seemed an endless quest to improve the fire department. When he introduced his Snorkel apparatus in 1958, he considered equipping all three of the fireboats with the unique articulating boom and bucket device. Since WWII, and throughout the Cold War, fire departments had a close alliance with the Office of Civil Defense. To fully use equipment provided by this federal agency, Quinn in 1957 proposed a civilian, all-volunteer auxiliary of his department. Quinn approached Chicago Civil Defense Corps Director Charles J. Johnson to organize the unit, which was called the Chicago Civil Defense Fire and Rescue Service. The service operated a wide variety of equipment, which was carried in six land units that responded to greater alarm fires and other emergencies. Included in its inventory was a 16-foot, jet-propelled fireboat donated by local Rodi Boat Company in 1960.

During the 1970s the unit was reorganized to become the Emergency Preparedness and Disaster Service. Much to the disappointment of its members, Chicago disbanded the service in 1989.

The Chicago Fire Department had high-pressure turret wagons for use in conjunction with land apparatus and fireboats since the turn of the century. However, during Quinn's twenty-one-year span of leadership, these units were significantly improved by the department shop as high-pressure monitor systems mated with various commercial vehicle chassis. Capable of discharging enormous volumes of water, these units culminated during the 1970s with two behemoth rigs named Big John and Big Mo. As was the practice in many cities, the CFD also detailed trucks for the sole purpose of carrying large-diameter hose either to be stretched inland from fireboats or from pumper apparatus to turret wagons. These rigs were traditionally called hose wagons.

Demonstrations at the Chicago Post Office during the 1962 Fire Prevention Week included land apparatus, the fireboat *Fred A. Busse*, and two of Commissioner Quinn's innovations: the Snorkel truck and the jet fireboat, a 24-foot, pump-equipped Buehler Turbocraft. Courtesy of Curt Elie Collection.

The Chicago Fire Department was always willing to show citizens what they were getting for their tax dollar. The fireboat J4, a 24-foot Buehler Turbocraft, was displayed at a Chicago boat show during the early 1960s. The display showed to good advantage the vessel's jet ports, discharge outlets on the transom, and monitors. Author's Collection.

The Chicago Fire Department's third and fourth deliveries of Buehler Turbocraft, designated apparatus 6-8-3 and 6-8-4, complete a test run on the Chicago River in late April 1962. Their shop codes were G-184 and G-185, respectively. A firefighter is positioned in the bow hatch to operate the bow monitor. The boats could reach speeds in excess of 40 mph. Author's Collection.

Chicago Fire Department divers secure a capsized boat to their J boat. Speeds of more than 40 mph allowed the small, waterjet-powered craft to arrive quickly at emergencies. Author's Collection.

Chicago's big fireboats frequently saw use as floating pumping stations for land apparatus, some of which were unique configurations. High Pressure No. 3 was a 1956 International truck with a specialized hose body mounting a monitor. Throughout much of the department's history, land apparatus traditionally featured red and green lights, mimicking those required on marine vessels. Courtesy of Chuck Madderom.

After the St. Lawrence Seaway—which stretched 189 miles between Montreal and Lake Ontario—opened to navigation in 1959, allowing shipping from the Atlantic Ocean to the Great Lakes, Commissioner Quinn took immediate steps to bolster fire protection in ports and on Lake Michigan. His response included four small jetboats to support the department's three large fireboats. These highly maneuverable boats were coming of age and promised faster responses and less operating costs than massive fireboats. The boats reached speeds topping 45 mph and could pump 3,000 gpm. A primary source for such boats lay not far from Chicago in Indianapolis, where Buehler Turbocraft was turning out jet-powered boats.

The first jetboat was an 18-footer that went in service in 1961, followed by three additional 24-foot Buehler boats, one in September and two in May 1962. A 24-foot, welded aluminum boat built by General Marine was delivered in 1966. These boats were designated CFD unit numbers 6-8-1 through 6-8-5, and assigned shop codes G-182 through G-186. Firefighters manning the jetboats were qualified scuba

Versatility was a byword in the Chicago Fire Department Air-Sea Rescue Unit, which made wide use of aircraft and fireboats, often in concert. This was the first of three Bell 47 helicopters operated by the unit. Litters were attached to the float gear of this Model 47G-4 in 1967. Like military rescue helicopters, fire department helicopters often flew with divers. Author's Collection.

divers. During their first full year of operation, the five boats collectively logged 119 responses. During the late 1960s, as the jetboats were phased out, they were replaced by larger, conventionally powered boats. The fleet included a 20-foot boat, three 35-footers, a 24-footer, and two former US Coast Guard vessels, most of which were used cooperatively by the Chicago Police and Fire Departments.

The Chicago Fire Department was no exception to the 1960s trend in using helicopters for special operations. Use of a leased Bell 47 helicopter in 1964 led to formation of Chicago's Air-Sea Rescue Unit in 1965, complete with its own Bell 47. The aircraft, and successive models, worked closely with fireboats during search and rescue operations. Like the fleet of small boats, the choppers eventually were operated cooperatively with the police department. Chicago's modern Air-Sea Rescue Unit operates a pair of Bell 412 helicopters.

Throughout the 1960s and 1970s, the big boats *Medill, Schlaeger*, and *Busse* performed admirably, often at fires that required their combined firepower. On January 16, 1967, one of those blazes devoured the gargantuan, supposedly fireproof McCormick

When Chicago Turret Wagon 6-7-2 was organized in September 1966, it used this 1962 Willys Jeep modified by the shop to deliver 4,000 gpm. Author's Collection.

King among the heavy hitters often used in conjunction with Chicago fireboats was *Big John*, named after its designer, John F. Plant. The unit was an example of the ingenuity and skill of fire department shop personnel across the country. Assigned shop code G-156, shop personnel mated a 1952 IHC M65 five-ton chassis and cab with two Stang high-pressure monitors capable of discharging 15,000 gpm. The unit was organized on May 12, 1970, and housed with Engine 42 at 55 Illinois Street. Author's Collection.

Similar to *Big John*, *Big Mo* was also shop-built with dual monitors fed by multiple hose inlets. Assigned shop code G-248, the rig was built on a 1953 IHC truck. In Chicago Fire Department tradition, the unit was painted red with a black cab roof. Author's Collection.

Drawing from their experience building high-pressure apparatus, Chicago Fire Department shop personnel fabricated more compact, yet powerful and efficient units. Turret Wagon 6-7-5 mounted a pair of Stang monitors on a 1975 Mack MB cab and chassis. Courtesy of Chuck Madderom.

There was no shortage of manpower aboard the *Busse* to retrieve an auto that went off the Archer Avenue Bridge in July 1964. Divers secured lines using the mooring line on the capstan and another through a pulley on the bow davit. Draft marks on the *Busse's* bow indicate it is drafting 7 feet of water. Courtesy of the Curt Elie Collection.

During the late 1960s and early 1970s, three 20-foot aluminum boats and two 35-foot all-steel boats were used cooperatively by Chicago's fire and police departments. The partner to this 35-footer wore the number 2. Courtesy of Bill Friedrich Collection.

With sixteen people crammed aboard, three of whom are divers, apparatus number 6-8-3, a 42-foot, diesel-powered Peterson boat, departs Burnham Harbor for the site where a twin-engine plane crashed during April 1977. Author's Collection.

Place, the largest convention center in the nation. Since the ten-acre site was built along the lakefront, the *Joseph Medill* was part of the full first-alarm assignment for Box 798. Ironically, the *Medill* was named after the former Chicago mayor, the grandfather of Robert R. McCormick, for whom the convention center was named.

The five-alarm-plus inferno at McCormick Place saw the three fireboats not only engaged in battle with mountainous flames, but drafting from Lake Michigan to supply water to land companies. The *Medill* took up a position on the east side of the complex; the *Victor L. Schlaeger* and the *Fred A. Busse* took up positions at the center's north and south ends, respectively. Despite the monumental efforts of firefighters working in frigid temperatures, it would take nearly ten hours to bring the blaze under control.

The crew of this 30-foot UTB Mark I conducts a body search in July 1974. The former Coast Guard utility boat was diesel-powered, having a top speed of 22 mph. From 1954 to 1983, Coast Guard Yard Curtis Bay, Maryland, built 222 of the boats in three models: the Mk I was constructed of steel, the Mk II wood, and the Mk III plastic. Author's Collection.

Children of a Chicago day camp in 1970 enjoy a demonstration by fire department Lieutenant Joseph Volpi. Assigned to the fireboat *Joseph Medill*, Volpi explains the operation of one of the Medill's powerful monitors. Author's Collection.

In early 1965, when Chicago fire officials formulated plans for an air-sea rescue unit teaming leased helicopters with fireboats, they conducted tests with the local firm Helicopter Air Lift. In March, this Bell 47 and the fireboat *Medill* simulated rescues from vessels. Courtesy of Curt Elie Collection.

A vessel taken into dry dock maneuvers over pre-set support timbers in a flooded dock and the water is then pumped out. The rail system seen here had the fireboat *Busse* maneuver over the submerged rail carriage, which was then pulled from the water. Courtesy of David S. Berger.

Boat 4 was a former Coast Guard 40-foot UTB, 236 of which were built by Coast Guard Yard Curtis Bay from 1950 to 1966. Displacing 11 tons and having a steel hull with aluminum superstructure and deck, the boat handled well in choppy water. Two 190 hp diesel engines drove twin screws. The Chicago Fire Department acquired the UTB after the type became surplus in 1978. The city flag was displayed diagonally on the hull. Courtesy of Bill Friedrich.

Dated bridges not closing properly in high temperatures has been a persistent problem in some cities. Here, the *Medill's* aft monitor is used to cool the ironwork of the Ohio Street Bridge over the Chicago River in 1989. Chicago's modern fireboat, the *Wheatley*, continues to address the problem of stuck bridges. Author's Collection.

Relics of a bygone era, Engine 37, *Medill*; Engine 41, *Busse*; and Engine 58, *Schlaeger* berthed at Lake Calumet during the 1990s. A fire department repairman often was detailed to the site to maintain the boats. The *Schlaeger* was kept at the filtration plant during summer months and at Navy Pier in winter, until replaced in 2010. Author's Collection.

Given the ever-looming budget constraints and the age of Chicago's trio of big fireboats, it was only a matter of time before they were earmarked for replacement by smaller, more cost-efficient vessels, especially in terms of manpower. The administration of Mayor Jane Byrne retired the *Busse* in 1981 after forty-four years of service at Navy Pier. On August 15, 1986, fire commissioner Louis Galante ordered the *Schlaeger,* stationed at Lake Calumet on Chicago's far southeast side, removed from active duty and placed on standby status with a crew of five; her assigned manpower had been four pilots and thirteen crewmen. The *Medill* was also relegated to Calumet after its years of service on the Chicago River. The pair took up berths next to the *Busse,* which became an excursion boat at Door County, Wisconsin. The *Medill* went to Algoma, Wisconsin Harbor, for storage on the Ahnapee River, and ended up on land at Escanaba, Michigan, earmarked for scuttling as a dive site off Algoma.

In retirement, the fireboat *Joseph Medill* was relegated to storage on land at Escanaba, Michigan. No longer in the hands of faithful, attentive crew, it is at the mercy of the elements. It is seen here in August 2012; its owners were denied it use as an artificial reef when it was deemed environmentally unsafe. Courtesy of Charles Pomazal.

The void left by the big boats, in part, eventually was filled in 2009 by the fireboat *F. F. Eugene Blackmon*. The "fast rescue boat" was named after a Chicago firefighter/scuba diver who drowned in May 1998 while trying to save two drowning victims in the Calumet River. Lake Assault Boats of Superior, Wisconsin, constructed its hull number LC33 of welded aluminum. Although a far cry from the *Medill* it replaced near Navy Pier, the *Blackmon* typifies the ability of smaller boats to deliver more. Within its 33-foot length and 10-foot beam are twin Honda 225 hp outboards, giving it a top speed of 45 mph. A Hale 650B pump in direct drive to a 5.7 liter GM 300 hp engine provides 1,500 gpm at 150 psi.

Chicago's fast rescue boat 6-8-8, named *F. F. Eugene Blackmon*, was built by Lake Assault Boats of Superior, Wisconsin, in 2008, and began service the following year. A unique feature of the 32-foot vessel is a 5½-foot-wide hydraulic bow door. A pair of 225 hp outboards allows a top speed of 45 mph, with 1,500 gpm possible through a 300 hp V-8 Hale pump. Courtesy of Lake Assault Boats.

By fire commissioner Robert S. Hoff's order, the *Victor L. Schlager* finally was removed from service on May 3, 2011, when it was replaced by the ultra-modern fireboat *Christopher Wheatley*. Named in honor of Chicago firefighter Christopher Wheatley who lost his life in August 2010, the steel fireboat with aluminum superstructure was designed specifically to function in Chicago's environment. That meant navigating shallow rivers, passing under low bridges, and breaking up to 16 inches of ice to remain operational year-round. The *Wheatley* was designed by Canadian Naval Architects Robert Allan Ltd., and is one of several fireboats designed by the firm and built in Canada under the Canada-US Free Trade Agreement. Hike Metal Products, coincidentally of Wheatley, Ontario, Canada, built the *Wheatley* as a Ranger 2700 Class fireboat, which complies with *NFPA Type III Standards for Fireboats*; Type III must have a minimum pump rating of 4,500 gpm through two pumps and carry enough fuel to stay on station for eight hours. Built as hull number 178, the vessel measures 90 feet with a 25-foot beam. Its height of only 16 feet above the waterline is achieved by hydraulically lowering the aft 30-foot monitor tower and mast. The boat was carefully designed to balance a draft of 7½ feet and overhead clearance, or air draft, from bridges.

After arriving in Chicago in April 2011, the fireboat *Christopher Wheatley* was organized as Engine 2 and berthed at the Chicago Marine Safety Station with the *Eugene Blackmon*. Although a day boat, the *Wheatley* has complete facilities for a crew of five, which increases to ten when fighting fires. Stored in the large after-house are life rafts and life jackets for 800 people. Courtesy of Gordon J. Nord Jr.

Details of the emblem displayed on the superstructure of the *Christopher Wheatley*. Courtesy of Gordon J. Nord Jr.

Chicago Fire Department officials used the Robert Allan-designed Baltimore fireboat built by Hike Metal Products as a guide; the lines are similar but the *Wheatley* had more restrictions—most notably height and higher pump rating—than Baltimore's boat built four years earlier. Darren Hass, Robert Allan's project engineer for the *Wheatley* explained in the company's May 2011 newsletter:

> We took a more conventional ice-breaking tug that we sort of scaled and stretched and attached to a more conventional twin screw, which is what the fireboat's hull is based on. It is much better suited than the hull for the Baltimore fireboat. The biggest challenge was sufficient space for clearance of the propellers.

Powering the *Wheatley* are twin 1,600 hp Caterpillar C32 diesel engines driving Rolls Royce Nibrel propellers strengthened for use in ice. Two caterpillar 1,000 hp diesel engines drive two pumps that provide 15,000 gpm at 150 psi. Three Stang monitors are mounted on the bow and one in the aft tower. The fireboat's 1,000 gallons of foam was a federal government requirement of the port security grant that funded more than half of the vessel's cost. With the ability to function as a fireboat, command center, rescue platform, and pumping station, the *Christopher Wheatley* is a sound investment for a city that at one time in its long history operated nine fireboats.

Chicago's *Schlaeger* demonstrates the practicality of low-profile fireboats in cities with numerous street bridges spanning major rivers. Courtesy of Gordon J. Nord Jr.

The *Schlaeger*, still looking good and going strong in 2008, nearly six decades after it was launched. The flag of the City of Chicago is displayed throughout its fire and police departments. The white field represents the city's three geographical areas, blue stripes signify the city's waters, and four stars mark historical events: Fort Dearborn, the Great Chicago Fire, the World's Columbian Exposition, and the Century of Progress. Courtesy of Gordon J. Nord Jr.

Milwaukee

Chicago's influence in fireboat design would become apparent 90 miles north of the Windy City, along the Lake Michigan shoreline, at Milwaukee, Wisconsin, which would lay claim to operating one of inland America's largest freshwater fireboat fleets. By the mid-nineteenth century, commerce flourished along Milwaukee's three rivers: the Milwaukee, the Menomonee, and the Kinnickinnic. Eventually, riverfronts became so crowded with railroad yards, lumberyards, and warehouses of heavy industry that fire companies could no longer depend on the rivers as a water supply. Fire Chief James Foley's proposal for fitting a steam pump aboard a river tug to fight blazes in the river valleys would prove unnecessary after the fire loss along Milwaukee's rivers in 1887 convinced city administrators to purchase a fireboat. Chief Foley recommended a first-class vessel similar to Chicago's new *Geyser*, which led to Captain Frederick C. Starke, whose family had marine interests in Milwaukee. Starke's father was credited with building Milwaukee's first tug in 1858. The captain's fireboat design became reality at Reiboldt, Wolter & Co. of Sheboygan, farther north along Lake Michigan's shoreline. Foremost among the vessel's machinery, which was built by Chas. F. Elmes Engineering Works of Chicago, was a 25-ton boiler built by Thomas L. McGregor, the largest boiler ever made in Milwaukee. Adding to her 245-ton displacement was a white oak hull sheathed with iron plating for ice-breaking.

Stretching nearly 107 feet, with a 24-foot, 6-inch beam and drafting 11 feet, it was capable of 5,000 gpm through vertical duplex pumps at 120 pounds steam pressure. During tests, its single monitor threw a 4-inch stream more than 400 feet.

After being christened *Cataract* by Milwaukee Mayor Thomas H. Brown on September 4, 1889, it slid the ways, finally arriving in Milwaukee during late October. The *Cataract's* crew consisted of pilot, marine engineer, and stoker, plus two officers and three firefighters. A portion of the fireboat's manpower was diverted to a boat wagon placed in service to overcome the difficulty of stretching the *Cataract's* 3,000 feet of hose inland. The two-horse wagon responded to alarms with the fireboat, its hose often laid prior to the boat's arrival.

In view of the *Cataract's* success, and with plenty of work for a second boat, Chief Foley's request for a similar fireboat was approved. Commissioned on December 1, 1893, the new *James Foley* was designed by Milwaukee's John Conrad, built by Reiboldt & Wolter, with Elmes machinery and a boiler by Chicago's John Mohr & Sons. Its 600 hp engine drove pumps that produced 4,500 gpm. Construction of the fireboat pipeline begun that year included special hydrants that incorporated an electrical signal system for communicating with the boat pilot. To harness extreme pressures pushed through large hose lines on land, the shop devised massive monitors with 6-foot anchor legs called street jacks.

Crew and visitors line the deck of Milwaukee's first fireboat, the *Cataract*, Engine Co. 15, at Jones Island shortly after its arrival in 1889. Although capable of delivering 5,000 gallons of water per minute, the *Cataract* mounted a single monitor, a common practice during the period. Metal plating that reinforced the bow for ice-breaking is visible on the bow. Courtesy of Milwaukee Fire Historical Society.

Since laying large-diameter heavy hose from fireboats to the fire was slow and back-breaking, the Milwaukee Fire Department shop modified chemical wagons to do the job. This was one of two chemical wagons converted in 1896 to carry 600 feet of 3½-inch hose. Chemical hose for the tanks below was stored in the wire basket between the wheels. Intricate detailing of the wheels spoke to the skill of shop personnel, who became known for scratch-building Milwaukee fire apparatus. Courtesy of Milwaukee Fire Historical Society.

Milwaukee's second fireboat was named after *James Foley*, the fire chief then in command and responsible for the boat's acquisition, along with the *Cataract*. Political infighting in 1903 over naming Milwaukee's fireboats led Foley to declare that all fireboats would be given numbers, including his own. So the James Foley became simply MFD No. 17. Courtesy of Milwaukee Fire Historical Society.

Since delays in opening numerous bridges, along with heavy river traffic, impeded fireboat responses to heavy industry along upper reaches of the Milwaukee River, a third boat went in service on April 1, 1897. Like its predecessors, the vessel came from the Reiboldt & Wolter yard with a pump rating of 4,500 gpm. Named the *August F. Janssen,* in honor of a chief killed at a blaze three years earlier, the boat in December 1898 pumped at a fire for ninety-seven hours, forty minutes, then said to be a world record.

Hard work on the polluted rivers and during bitter winters took its toll on the wooden fireboats, so in 1901 a new steel fireboat was approved. The 118-foot fireboat was of Chicago origin, having been designed by Chicago naval architect W. J. Wood, and built by Shipowners Dry Dock Co. in Chicago with a Mohr boiler and engines from Chicago Shipbuilding Co. Only its three 3,000-gpm pumps were not native, having come from the American Fire Engine Co. of Seneca Falls, New York. Before the new fireboat was

Since bridge-opening delays and heavy boat traffic impeded fireboat response to the northern reaches of the Milwaukee River, Chief Foley secured the go-ahead for a third fireboat, which began service in April 1897. Named the *August F. Janssen* in honor of a chief killed at a major blaze in 1894, it was renamed *MFD No. 23* after the 1903 fireboat name debate. The *Janssen* pumped for ninety-eight continuous hours at a fire in December 1898, then said to be a world's record. Courtesy of Milwaukee Fire Historical Society.

Milwaukee Fire Department boat wagon numbers corresponded to fireboats to which they were assigned. Boat Wagon 23, seen here in 1911 with a newly built wagon, was stationed near *MFD No. 23's* downtown berth. Eventually, horses reluctantly surrendered their place on the reins to motorized hose cars. Courtesy of Milwaukee Fire Historical Society.

With fresh coal shoveled into its firebox to power through ice on the Burnham Canal in 1936, *MFD No. 15* puts out as much smoke as some fires to which it responded. Ice-breaking on Milwaukee's waterways in January was on a par with firefighting for fireboat crews. The 106-foot *MFD-15* was Milwaukee's fourth fireboat and its first built of steel. After retirement in 1952, it embarked on a new career with Great Lakes Towing in the Northeastern US. Courtesy of Milwaukee Fire Historical Society.

commissioned on November 3, 1903, Chief Foley eliminated controversy in city hall over naming the boat by changing all fireboat names, including his own, to numbers. The *Foley* became *MFD-17,* the *Janssen* became the *MFD-23,* and the new boat was christened *MFD-15.* The *Cataract* was retired and its machinery installed in a steel hull at Manitowoc Dry Dock Company to become Milwaukee's fourth fireboat, which was christened simply *MFD-29.*

As Milwaukee added fireboats, the number of boat wagons increased, with chemical wagons converted to boat wagons, supplemented by larger,

The *MFD-29* teams with two sister fireboats to fight the Pabst Elevator B fire on October 28, 1909. It was commissioned on June 2, 1906, and served both northern and southern reaches of Milwaukee's rivers. The steel tower abaft the funnel was found impractical and was removed a few years after beginning service. Courtesy of Milwaukee Fire Historical Society.

shop-built wagons mounting monitors. Each boat was assigned a boat wagon.

Milwaukee's four fireboats gave good service, with three usually seen working at major fires; peak year was 1909, during which the boats responded to more than 500 alarms. By the early 1920s, however, a number of factors would change the face of marine firefighting in Cream City. Vessels grew old before their time and operating costs began to climb. Shipping had declined and industry moved inland. A major change in the fire department's work shift meant new hires, with reduction of the fireboat fleet eyed as a means of paying their wages. The first step in economizing was to retire the former *Janssen* and the *Foley.* Their replacement in 1922 was the floating powerhouse *Torrent,* which had served the Duluth & Iron Range Railway since it was launched at American Shipbuilding Co. of Lorain, Ohio, in 1910. Weighing 296 gross tons and measuring 110 feet with a 28-foot beam and drafting 15 feet, *Torrent's* twin turbine pumps produced 12,000 gpm.

The *Torrent* when it arrived in Milwaukee during October 1922. To replace the *MFD No. 17* and *MFD No. 23*, the floating powerhouse was purchased from the Duluth & Iron Range Railway. Weighing 296 gross tons and measuring 110 feet long, the all-steel *Torrent* would be Milwaukee's largest fireboat, capable of pumping 12,000 gpm. Courtesy of Milwaukee Fire Historical Society.

In July 1923 the *MFD-23 (Janssen)* was dismantled and sunk in Lake Michigan, while *MFD-17 (Foley)* was kept in reserve until May 1930, when it too was sunk. In keeping with its reputation for building fire apparatus from the ground up, the fire department shop in 1927 built three additional boat wagons, each carrying 2,000 feet of 3½-inch hose.

From 1932 until the post-WWII era, the *Torrent* and *MFD-15* protected Milwaukee's waterways, backed up by *MFD-29,* which went on the reserve list in 1932.

Research in 1937 showed that converting the boats to diesel power was not cost effective, and the outbreak of war in 1939 ruled out long-term plans to update, much less purchase, new fireboats. It wasn't until 1948 that city officials acknowledged Chief Edward E. Wischer's proposal to replace all three boats with a new diesel-powered vessel patterned after Chicago's *Busse.* Defoe Shipbuilding, which had built the *Busse,* got the job and began work on a design by Chicago naval architect A. M. Deering.

Named *Deluge,* the all-steel, low-profile fireboat was launched as Defoe's hull number 420 on July 5, 1949. Weighing 141 gross tons, it measured 96 feet, with a 23-foot beam and 6-foot, 6-inch draft. Four 375 hp diesels each drove a 3,000-gpm pump; the two aft engines also provided propulsion. Although much of the city's shoreline hazards had disappeared, and land-based firefighting capability had improved vastly, the *Deluge* was good insurance. Despite the potential for disaster that remained—which some felt warranted a second fireboat—the logbook of the fireboat would record few major fires during its first two decades of service. Nevertheless, it and crew remained vigilant, and with its bow reinforced for ice-breaking, rivers remained open and floods were averted. Support apparatus too was kept at the ready, with two high-pressure trucks with monitors built during the late 1950s to supply the *Deluge* and fireboat pipelines with 3½-inch hose.

Besides regular duties of firefighting and ice-breaking, fireboat crews periodically were summoned to pull vehicles, bodies, and cattle from Milwaukee waterways. In 1956, *MFD No. 29*'s crew rescues a cow that escaped from a nearby stockyard near Canal Street in the industrial valley. Courtesy of Milwaukee Fire Historical Society.

With its steam turbines pushing smoke from its twin stacks, the *Torrent* aims its monitors at a coal yard fire, while also supplying hoses to land companies. It served twenty-seven years before being retired in 1949, and was scrapped the following year. Courtesy of Milwaukee Fire Historical Society.

Despite the *Deluge's* low profile, it could not pass under all of Milwaukee's street bridges, some of which were without operators due to manpower shortages, which caused serious delays. Compounding the problem was city hall's declaration that northern reaches of the river had become non-navigable. The success that the nearby Chicago Fire Department was having with its new small jetboats hadn't gone unnoticed by Chief James R. Moher and staff. Nor was it lost on them that these small boats, besides having the ability to pass beneath any bridge, were fast and maneuverable, and required less manpower and maintenance than large fireboats. After Milwaukee politicians in November 1962 approved the purchase of a 27-foot jetboat that could pump 1,000 gpm, the next stop was Rodi Outboard Center of Chicago, dealer for Fire and Rescue Boats.

The two monitors platform-mounted on *Deluge*'s bow originally were in a tub surround. Another monitor was on the stern and a fourth was in a tower that elevated 24 feet above the waterline. The boat's pristine condition reflected the pride of fireboat crews. Author's Collection.

Milwaukee's 12,000-gpm *Deluge*, in its original color scheme, breaks ice on the Milwaukee River in 1962. It was one of two fireboats built by Defoe Shipbuilding of Bay City, Michigan. The first was Chicago's *Busse* built in 1937. Courtesy of Chuck Madderom.

The *Deluge* pumps into a fireboat pipeline in 1966 to deliver power inland to companies fighting a three-alarm fire. By the 1940s, Milwaukee had thirteen miles of high-pressure fireboat pipelines laid, servicing nearly 300 boat hydrants. The system's 8- and 10-inch-diameter pipes were kept dry to prevent freezing. Author's Collection.

Without its radar mast, *Deluge*'s height above the waterline was only 15 feet, allowing it to pass under Milwaukee's many bridges. Large rectangular panels on the deckhouse fore and aft are doors providing access to discharge manifolds. Seen here in 1973, this was *Deluge's* second color scheme. Courtesy of Chuck Madderom.

The origin of Milwaukee's jetboat dates to the late 1950s when Chris-Craft Corporation acquired Roamer Boat Company of Holland, Michigan, renaming it Roamer Steel Boats Division. After launching a new line of 35- and 42-foot welded steel Roamers, the boats grew in popularity, and by 1961 the firm offered twenty-three models ranging from 31 feet to 56 feet. During 1962, the Roamer Steel Boats Division opened its line of aluminum boats, beginning with 27-foot cruisers known as Silver Comets and Custom Comets, 170 and fifty of which were built, respectively. A "Red Comet" model was offered, which was a fire-rescue modification produced only between 1962 and 1964. These boats, and eventually all Roamers, became rarities since production was transferred from Holland, Michigan, to Pompano Beach, Florida, where the line was ended in 1979. With a basic weight of 6,400 pounds, the Red Comet had a 10-foot beam, 2-foot draft, and an aft freeboard of 3 feet. Although the standard engine offered by Chris-Craft for its Roamers was a 185 hp GM Model 283, gasoline-powered V-8, Milwaukee's Red Comet was jet powered. Jet power may have been revolutionizing the boat industry, but jet intakes clogged by river pollution proved problematic, prompting department mechanics to devise an alternative suction intake.

The Wisconsin Avenue Bridge in the background was typical of Milwaukee's low bridges that proved troublesome for fireboats, necessitating tests of small fireboats. This waterjet-powered boat being demonstrated to officials on July 19, 1962, would join the *Deluge* at its Water Street berth. Courtesy of Milwaukee Fire Historical Society.

The Milwaukee Fire Department's Red Comet, which shared berthing space with the *Deluge* at the city's first fireboat station at Water Street, was designated *Engine 42,* later *Roamer,* and finally *Fireboat 2,* serving only during summer months. Both boats got their share of work at major fires, with the *Deluge* often

Shortly after an advertising tour that included this fire prevention week parade in Chicago in October 1962, the Milwaukee Fire Department claimed the 27-foot Roamer as its second fireboat. Author's Collection.

Milwaukee's Fireboat 2, first labeled Engine 42, after outfitting in 1963. Rated at 1,000 gpm, it mounted monitors on stern and bow, with the latter operated from a bow hatch. Courtesy of Gerrit Madderom.

pumping for more than eight hours. The 1970s saw an increase in the boat's workload, and it was during one of these blazes that the *Deluge* broke the old *Janssen's* seventy-seven-year record for farthest distance pumping water inland. At the five-alarm Jefferson Street School fire in October 1978, the *Deluge* pushed water through 4,500 feet of hose.

Despite the impeccable condition of the *Deluge,* the addition of an onboard foam system in 1978, and a new high-volume hose truck, inflation caught up with the vessel, first trimming her crew to only pilot and engineer, and finally placing her out of service, along with *Fireboat 2,* in October 1984.

With the loss of the two fireboats inevitable, plans had been made to ensure that Milwaukee was not without waterfront protection. Since staffing a fireboat was the largest expenditure, careful planning was necessary to come up with a suitable, cost-effective means of marine firefighting. Success with Miami's military surplus LARC fireboat conversion spearheaded by Louis DeChime had turned heads, compelling chief William Stamm and staff to have a closer look. What they saw convinced them that an amphibious fireboat was the answer to the manpower issue. Thus, Ladder Tower's first LARC conversion premiered as Milwaukee's replacement for its two fireboats.

DeChime had sold LTI's president on the idea of a modified version of Miami's LARC fireboat that used conventional pumps to deliver 2,000 gpm through a forward monitor and two 1,000-gpm bow monitors, all three of which were electro-hydraulically controlled from the pilothouse.

Wearing a new color scheme, Fireboat 2 in 1973 after conversion from waterjet to conventional screw drive, and with an external suction to bypass the hull intake, which often clogged on the then-polluted river. Courtesy of John A. Calderone.

The 1960s vintage LARC-V-turned-fireboat was delivered in mid-October 1984 and went in service on November 4. It was housed initially with Engine Company 12, and then with Engine Company 3, both near Lake Michigan's shoreline. Engine Co. 3 was manned by members of the department's Dive Rescue Team, which also manned the LARC, thereby resolving the age-old problem of staffing costs for fireboats. Also beneficial was a significantly shortened response time, thanks to the amphibian's ability to be driven directly to the water, versus the time required for a land company to drive to, and launch, a boat.

Milwaukee's fireboat converted from a military LARC V displays its amphibious capability. Two 1,000-gpm bow monitors were electro-hydraulically operated from the pilothouse, while the 2,000-gpm monitor abaft the pilothouse was mounted atop a vertical pipeway. Author's Collection.

Weighing in at nearly 16 tons, the amphibian was powered by a 300 hp Cummins V-8 diesel engine, while its 2,500-gpm output—which was actually 2,950 gpm—was achieved with installation of two American (later Hale) two-stage pumps powered by a 492 hp Detroit eight-cylinder diesel engine. Installation of the more powerful NASA-developed Firefly pump unit was ruled out since it was cost prohibitive; the Firefly's price tag was nearly equal to the cost of the fireboat. Other features included an automatic transmission and hydraulic disc brakes. Engine 3's quarters easily accommodated the LARC's 35-foot length, 10-foot beam, and 11-foot height. Drafting 4 feet, its height in the water was a mere 7 feet.

The firm Ladder Towers used an ample amount of diamond plate to convert a LARC V to fireboat. Two single-stage, 1,250-gpm fire pumps were mounted at the center of the fireboat, connected to a common transmission. Courtesy of Bob Milnes.

Besides providing superior traction off-road, the massive tires of the amphibious fireboat aided stability in the water. Immediately aft of the suction boxes mounted amidships on the hull were conventional hydrant inlets, allowing the boat to function as a land pumper. Courtesy of Bob Milnes.

Milwaukee's amphibious fireboat uses its forward monitors to close with a four-alarm fire at the P & V Atlas Industrial Center Complex on September 27, 1988. Courtesy of Bob Milnes.

When unique fireboats no longer churned Milwaukee waterways, the city purchased this 1994 Boston Whaler Vigilant rated at 1,000 gpm. Named *Phoenix*, the 32-footer mainly supported the department's dive rescue team, whose equipment is displayed with the *Phoenix*. Author's Collection.

The Milwaukee Fire Department in 2007 joined the long list of customers for fire-rescue vessels built by Canada's MetalCraft Marine. The 36-foot Firestorm is rated at 5,000 gpm. Two years after the boat arrived, it was christened *Trident*. Courtesy of Bill Friedrich.

Since the amphibian, built in 1967, was already outdated at the time of its conversion—some drives and related parts came from construction equipment—its days were numbered. When the expected high repair costs and scarcity of parts spelled its end, some felt the city was lucky to have gotten nine years of service from it. The mechanic, along with members of the firefighting division, who worked directly with the unit, was quick to express satisfaction with the concept. The amphibian's demise may have been premature since plans had already been made for remanufacture with new components, even newly built aluminum hulls, should the supply of military surplus LARCs be depleted. But Milwaukee's LARC fireboat was retired in December 1993 and replaced by the less complex, although much smaller and less powerful, *Phoenix,* a 32-foot Boston Whaler Vigilant, rated at 1,000 gpm. The *Phoenix* would go it alone until mid-2007 when Milwaukee joined the long list of customers taking delivery of fire-rescue boats turned out by MetalCraft Marine of Ontario, Canada. Two years of service would pass before director of Milwaukee's Harley-Davidson Museum, Stacy Watson, christened the boat *Trident.* Built as part of MetalCraft Marine's Firestorm line, the 36-foot welded aluminum boat can travel at more than 40 mph, thanks to a pair of 540 hp Cummins diesel engines driving Hamilton jets. Twin Darley 1,750-gpm pumps at 150 psi provide 5,000 gpm delivered through five monitors and dual discharge outlets. In addition to a foam tank, the *Trident* features night vision and thermal vision side scan sonar technology.

Milwaukee's High Pressure 1 was built by Marion Body on a 1957 FWD chassis and cab. Carrying 2,000 feet of 3½-inch hose and mounting two monitors, the unit was designed to operate in conjunction with the fireboat, or the city's high pressure water main. Quartered at the main downtown station, High Pressure 1 responded on all alarms with the fireboat. Courtesy of Gerrit Madderom.

Duluth-Superior

At the far northern reaches of the Great Lakes, the largest US city on Lake Superior, Duluth, Minnesota, was making its mark early in its history. So lucrative was the lumber, mining, and railroad industry in the region that Duluth in 1870 was the fastest growing city in the nation. In fact, at the beginning of the twentieth century, shipping tonnage at Duluth's port surpassed that of New York and Chicago. Such rapid growth caused concern about the potential for fire along miles of railroad trestle, wooden docks, industrial complexes, and lumber stockpiles. Fortunately, a number of resources came together to provide firefighting capability along the waterfront of the harbor shared by Duluth and Superior, Wisconsin, which included the city of Two Harbors, twenty miles northeast of Duluth.

Milwaukee's pair of 1950s vintage high-pressure trucks was replaced in 1980 by this unique large-diameter hose unit. The large hose reel harked back to the days when massive reels appeared on the decks of fireboats. The unit's twenty sections of 5-inch-diameter hose was laid and retrieved by a low-speed motor. Barkow Body built the 5-ton unit using a 1979 Ford F350 truck chassis and cab. Courtesy of Gerrit Madderom.

When *MFD No. 15* was sold to Great Lakes Towing in 1953, it underwent an amazing transformation when rebuilt with modern superstructures, and its steam power converted to 1,750 hp diesel-electric. It was re-christened *Laurence C. Turner* on June 2, 1954, and later *Ohio*. Paul Csizmadia, Spec3 Photography.

Milwaukee's former fireboat built in 1903, as a rebuilt tug and now more than a century old, powers through heavy ice on the Cuyahoga River in 2010. Paul Csizmadia, Spec3 Photography.

Built in 1895 and 1896 by Cleveland Shipbuilding Company for the Duluth Missabe and Iron railroad, the *Edna G.* called Two Harbors, Minnesota its hailing port for nearly its entire 86-year career on the water. The huge 3,000 gpm monitor on its aft deck saw frequent use for firefighting and for washing docks after ore boats loaded. Author's Collection.

One of the first historic workboats in the region was the tug *EDNA G.,* one of the most powerful tugs of its time. Construction of the steel-hull vessel was begun by Cleveland Shipbuilding in 1895 for the Duluth and Iron Range Railroad. Hull number 25 was named after the railroad president's daughter, Edna Greatsinger, and slid the ways on April 4, 1896. Measuring 110 feet long, with a 23-foot beam and drawing nearly 16 feet of water, its bunkers held 70 tons of coal to keep power on its 1,000 hp double-expansion steam engine. When it was under way, its fireman fed a dozen large shovels full of coal into both sides of its boiler every ten minutes; average daily coal use was 6 tons. For firefighting, mounted aft on the deckhouse was a 3,000-gpm monitor. Except for two years during WWI, when the government outfitted it with a steel superstructure and pressed it into service on the East Coast, the *Edna G.* called Two Harbors its hailing port during its eighty-six-year career. When it was retired in 1981, it was the last coal-fired, steam-powered tug on the Great Lakes.

Among the tugs that called Lake Superior home was the *America,* a steel-hull 83-footer weighing 93 gross tons. It was built in 1897 by Buffalo's Union Dry Dock Co., for private ownership, and in 1916 went to Great Lakes Towing Co. Mounted on the bow of *America* and its sister tugs was a powerful monitor for waterfront fires.

The 93-foot tug *America* drew a crowd as it fought this waterfront blaze at Duluth during the early 1900s. It was built in 1897 as hull number 81 by Union Dry Dock of Buffalo for Great Lakes Towing Co. A sister tug lies alongside to assist. Courtesy of Hugh McKenzie/University of Wisconsin-Superior. Jim Dan Hill Library.

Despite additional protection of Lake Superior loading docks provided by US Navy-supplied fireboats and tugs with firefighting capabilities, more was needed, spurring the railroad to act. Officials of the Duluth and Iron Range Railroad went to American Shipbuilding Co. (AmShip) Lorain of Lorain, Ohio, which launched its hull number 364 in 1908. Shortly thereafter, the Duluth and Iron Range Railroad merged to become Duluth Missabe and Iron Range Railroad, nicknamed Missabe Roads. The all-steel, 110-foot vessel was named *William A. McGonagle.* During its twenty-seven years with the railroad, the *McGonagle* earned a reputation as a dependable fire tug with its ability to throw powerful streams great distances.

Next from AmShip Lorain came the aptly named *Torrent*, built specifically for Missabe Roads. With the ability to pump 12,000 gpm, the *Torrent* was king of the fire tugs that plied Lake Superior. It served until the last wooden docks were rebuilt of steel, and was then sold in 1922 to the Milwaukee Fire Department. By the 1950s, further decline in fireboat activity along the Iron Range reflected the decline in high-grade iron ore in the Duluth-Superior region.

Equipped with steam-turbine propulsion and pumping that enabled 12,000 gpm, the fire tug *William A. McGonagal* was technologically superior to most fireboats of its day. Here, the mighty *McGonagal* throws impressive streams at the Le Seuer Lumber Mill fire at Duluth on July 10, 1909. Courtesy of Hugh McKenzie/University of Wisconsin-Superior. Jim Dan Hill Library.

While visiting the Iron Range in 1912, representatives of US Steel line the superstructure of the Missabe Roads fire tug *Torrent* as it demonstrates its monitors in Two Harbors, Minnesota. Courtesy of the Dalke Collection/Minnesota Museum of Mining.

The *Torrent* slides the ways at American Shipbuilding Co. at Lorain, Ohio, in 1910. After outfitting, the 12,000-gpm *Torrent* reigned as king among the fire tugs on Lake Superior. Author's Collection.

The fire tug *Torrent* of the Duluth & Iron Range Railroad Co. in 1910. Besides its powerful bow monitor, it mounted one atop the pilothouse and one on the stern. Courtesy of Hugh McKenzie/University of Wisconsin-Superior. Jim Dan Hill Library.

Cleveland

On the southern edge of Lake Erie and encompassing the winding Cuyahoga River, Cleveland, Ohio, saw rapid growth after the Ohio and Erie Canal was completed in 1832, turning the city's port into a key link between the Ohio River and the Great Lakes. Two years later, Cleveland's first organized firefighting company was established. The department's history with fireboats, however, began fifty years later, when the first fireboat was among new equipment purchased in the wake of disastrous fires in the industrial region along the Cuyahoga called the Flats. And since Cleveland had become a manufacturing giant during the Industrial Age, fireboats would become a mainstay on the busy waterfront.

Cleveland's fireboat *John H. Farley* was the former *J. L Weatherley* launched on August 4, 1886. It emerged from rebuild in 1894 measuring 90 feet long and capable of pumping 3,360 gpm, compared to the *Weatherley*'s 2,400 gpm. Officials inspect the Farley in September 1934. Author's Collection.

Named after city official and prominent volunteer firefighter John L. Weatherley, the *J. L. Weatherley* was launched on August 4, 1886, by Cleveland's Excelsior Iron Works. At 66 gross tons, the iron-hulled fireboat measured 79 feet long and was capable of pumping 2,400 gpm. In 1894 it was rebuilt by Bell's Steam Engine Works of Buffalo to 109 gross tons, measuring 90 feet and capable of pumping 3,360 gpm. It emerged from rebuild renamed *John H. Farley,* after Cleveland's former mayor.

That same year, on June 1, the fireboat *Clevelander* was commissioned, having been built by the city's engine builder Thomas Manning, with its wooden 70-foot hull built by another firm. Weighing 82 gross tons, the *Clevelander* drew 8 feet, 6 inches of water and could pump 5,010 gpm. At the end of its firefighting career in April 1916, it was towed by the *Roumania* to Manitowoc, Wisconsin, where its engine was installed in Cleveland's new fireboat *George A. Wallace,* named after the city's fire chief from 1901 to 1931; *Clevelander's* hull was then burned.

Manitowoc Shipbuilding Company built the steel-hulled *Wallace* at 112 gross tons, 77 feet long, and having a beam of 20 feet, 6 inches, with a 12-foot draft; she could pump 5,000 gpm. The *Wallace* served the Cleveland Fire Department for twenty years before being converted in 1936 for tow work in Canada, where it was scrapped in 1953.

Cleveland's second fireboat was the 70-foot *Clevelander*, which enjoyed a thirty-two-year career before it went to Manitowoc, Wisconsin, to have its equipment installed in its replacement. Seen here in 1913, the *Clevelander* was overall dark green with tan bulwarks. Courtesy of Michael Schwartz Library, Cleveland State University.

The *John H. Farley* was retired in 1938. Rebuilt in Canada as a fire tug named *Fire Chief*, it sank at its dock in 1968, only to be raised and scrapped the following year.

For the next eight years, Cleveland relied on US Coast Guard vessels for fire protection of its waterfront. Never were their services more vital than during the disastrous East Ohio Gas Company explosions that occurred on October 20, 1944, destroying more than eighty buildings and claiming 130 lives.

Like many fire departments during the postwar period, Cleveland took advantage of the availability of military surplus equipment. In 1946 the Cleveland Fire Department acquired from government liquidation a former fishing vessel built in 1927, which the Coast Guard had pressed into war service after outfitting as a fireboat in 1942. Number CG-64022F began life as hull number 520A, a gasoline-powered fishing tug built by Great Lakes Engineering Works of Ashtabula, Ohio. Resuming its original name *Mavret H.* for fire service in Cleveland, the steel-hulled boat weighed 24 gross tons. It was 60 feet long, having a 14-foot beam and a 4 foot draft.

For nine years, the *Mavret H.* shared fire protection of Cleveland's waterways with *The Clevelander,* a steel-hulled fireboat built in 1954 by Paasch Marine Service of Erie, Pennsylvania. Rated at 6,000 gpm, *The Clevelander* displaced 28 tons and measured 58 feet, with a 16-foot beam and 6-foot draft. On retirement in 1963, the *Mavret H.* was converted for towing and remains current on the US vessel registry.

Cleveland fire officials again called on Paasch to build a fireboat, which would be renowned for its longevity. Named for Cleveland's forty-ninth mayor, the *Anthony J. Celebrezze* was delivered to Cleveland on October 20, 1961. The steel-hull, 6,000-gpm boat is 60 feet, 8 inches long, having a 16-foot beam and drawing 6 feet, 9 inches of water. *Celebrezze's* 13-foot overhead clearance allows passage under all but three of the city's ten river bridges. Powering the *Celebrezze* are four Detroit diesel engines; two rated at 215 hp for either propulsion at 17 knots, or to drive a pair of single-stage, 1,500-gpm pumps. The other two diesels, rated at 250 hp, drive two additional 1,500-gpm two-stage pumps. At 42 gross tons and with its strengthened steel bow, the *Celebrezze* makes quick work of ice during frigid Midwestern winters.

Cleveland would boast two capable fireboats for thirteen years until *The Clevelander* was retired in 1974 and converted for tug work.

Throughout its history, the Cleveland Fire Department quartered its fireboats at Station 21, first established in December 1894 on the Cuyahoga River's west side. A new Station 21 was established in 1923, which remains the home of the *Celebrezze.*

Cleveland's 6,000-gpm fireboat *Clevelander* undergoes sea trials in Erie Bay in September 1954. Courtesy of Michael Schwartz Library, Cleveland State University.

When this photo was taken of Cleveland's fireboat *Celebreeze* in 1988, it was responding to more than 300 calls annually. The *Celebreeze*, which was also designated Engine 21, carried an 8-foot aluminum boat aft of the funnel. Courtesy of Curt Elie.

In September 2012, the Cleveland Division of Fire placed in service a 27-foot Boston Whaler to enhance fire-rescue coverage of the city's lakefront and the Cuyahoga River.

Buffalo

Despite the strong, proud history of battleship-class fireboats of major cities across the nation, the oldest operational fireboat in the world is not where most would expect—it is on the Great Lakes, specifically, Buffalo, New York. Already well into its second century of service, Buffalo Fire Department's *Edward M. Cotter* survived a devastating explosion and fire, countless blazes, fierce winters, numerous upgrades, and political battles. But before the *Cotter*'s keel was laid, Buffalo's firefighters had already become seasoned in the use of fireboats. As had happened in many port cities, Buffalo's first fireboat was born of necessity to keep pace with the city's growth into a major industrial center. After continued growth and the boom in maritime commerce brought about establishment of a paid, professional fire department in 1880, the need for a fireboat for port protection was realized. Buffalo city officials turned to Scottish immigrant David Bell, a local engine builder, who had established Bell's Steam Engine Works. In 1887 Bell launched an iron-hulled fireboat that measured 78 feet with a 20-foot beam and 9-foot draft. A hefty vessel at 133 gross tons, it could pump 3,810 gpm. The fireboat began service as the *City of Buffalo*, but in 1903 was renamed *George R. Potter* (Engine 29) in honor of the fire commissioner. The *Potter* was augmented in 1893 by the *John M. Hutchinson* (Engine 23), also built by Bell. Weighing much less at 90 gross tons, the steel-hull boat was 75 feet long and pumped 5,000 gpm.

When Buffalo began work on an extensive high-pressure fireboat pipeline in 1898, plans were made for a fireboat having the combined power of the *Potter* and *Hutchinson* to pump into the pipeline at every major downtown fire, and double as an ice-breaker to keep open Buffalo's harbor and the city's water intake. Authors of the fire department's 1900 annual report could not have imagined the actuality of their words when they wrote, "When completed, this boat will be the equal, if not the superior, of anything of its kind afloat, and will give ample protection to the property along our waterfront for years to come." Crescent Shipyard of Elizabeth Port, New Jersey, got the job and wasted little time in building the fireboat. Her keel was laid in March 1900, and on September 5, the fire commissioner's daughter, Virginia Pearson, christened her *William S. Grattan* after Buffalo's first paid fire commissioner.

The *Grattan* (Engine 20) weighed 274 gross tons and was 118 feet long with a 24-foot beam and 11-foot draft. The lower portion of its steel hull was wrapped with a 1½-inch-thick belt of Swedish steel for ice-breaking. Its three double-acting steam pumps supplied water at 9,000 gpm to three monitors.

The *Grattan* served admirably for more than a quarter century until fate caught up with her, proving disastrous for boat and crew. On July 28, 1928, the vessel responded to the foot of Maurice Street where the oil barge *James F. Cahill* was moored, now fully aflame. The *Grattan* and land companies worked tirelessly to keep the blaze from spreading, but

After maintenance, Buffalo's 80-foot fireboat *George R. Potter* prepares to leave dry dock in March 1918. Water is being pumped into the basin to float the vessel clear of cribbing and strongbacks. It would then leave the dock under its own power. Author's Collection.

The *John M. Hutchinson*, Engine 23, was built in 1893 as Buffalo's second fireboat. Weighing 90 gross tons, it measured 75 feet long, with a 20-foot beam and 9-foot draft. The *Hutchinson's* firepower was concentrated on its foredeck, while a large hose reel occupied its afterdeck. Eight discharge outlets for hose lines were built into the broad face of the pilothouse. Author's Collection.

At the beginning of the twentieth century, Buffalo's *Edward M. Cotter* embarked on a long and arduous career as the *W. S. Grattan* seen here in dry dock. Twin funnels provided exhaust for two Babcock and Wilcox 10-by-9-foot water tube boilers. These gave the Grattan a speed of 15 mph and powered three double-acting pumps built by American Fire Engine Company. Besides the bow monitors, a third monitor was mounted on the stern. Author's Collection.

seventeen hours into the fire, burning oil drifted across the river to the oil tanker *B. B. McColl.* The tanker exploded, engulfing the *Grattan* in flames. The pilot and his crew of eight jumped into the water and swam through the flames to shore. All were injured and the chief engineer, too severely burned to hold a life-line, slipped beneath the surface and drowned. The boilers of the burning *Grattan* soon ran dry and exploded, leaving the vessel a burned out steel hull.

The fireboat *Potter,* which had been placed on reserve status in 1926, was placed back in service while city officials grappled over the *Grattan's* fate. Their options were rebuilding the fireboat, purchasing a new one, or purchasing the fire tug *McConagle* being offered by Duluth's Massabe Roads. With the Depression gripping the nation, prudence dictated buying the powerful, reasonably priced *McGonagle.* But Buffalo Fire Commissioner George W. Hedden stubbornly refused, stating, "We're not in the market for any stuff like that . . . If we buy any more tugs, we'll buy them new." And so it went, the officials embroiled in debate for two years until the decision was made in 1930 to rebuild the *Grattan.*

The contract went to Buffalo Dry Dock Co., which installed oil-burning boilers, raised its pilothouse one complete level, and installed a 22-foot steel tower with monitor at the aft end of the upper deck. A fourth monitor was mounted atop the pilothouse and a large hose reel was added to the fantail. Finally, on November 19, 1930, *W. S. Grattan* returned to its Ohio Street berth.

At the half-century mark, the *Grattan's* boilers were incapable of maintaining steam pressure to operate at full gpm output. After it was deemed less costly to convert to diesel power, in 1952 the *Grattan* went to Wisconsin's Sturgeon Bay Shipbuilding & Dry Dock Co. for the conversion and major rebuild, using plans drawn by A. M. Deering & Son Naval Architects.

Steam engines and boilers were replaced by four Caterpillar D-397 diesel V-Type 12-cylinder powerplants rated at 400 hp for continuous duty. All four engines drive four De Laval centrifugal pumps, each rated at 3,750 gpm at 160 psi. The pair of engines installed aft also provides propulsion, turning twin screws for increased maneuverability. The rebuild lightened the *Grattan* to 178 gross tons, reducing her draft to 7 feet forward and 11 feet aft.

After a half century of service, in 1952 the *Grattan* went to Sturgeon Bay, Wisconsin, for rebuild, which included conversion to diesel power and replacement of its fixed monitor tower with a hydraulically operated tower with sliding ladder. Low, streamlined dummy funnels housed exhausts of the newly installed diesels. So complete was the rebuild that the 15,000-gpm fireboat was renamed *Firefighter*. A year later it was renamed *Edward M. Cotter*. Courtesy of Richard Sikora.

Exterior modernization included lower "dummy" funnels to house the exhaust of new engines, and replacement of the fixed monitor tower with a hydraulically elevating tower with sliding ladder. A fifth monitor was added on the fantail and a pair of 230-gallon liquid foam tanks was installed below deck. So complete was the *W. S. Grattan's* rebuild that when it arrived back at Buffalo during late 1953 it had been renamed *Firefighter*. It was renamed again the following year—to *Edward M. Cotter* in honor of the firefighter and president of Firefighters Local 282 AFL-CIO who had recently died.

Unique to the *Cotter's* long history was its night crossing of the international line in response to a plea for help from officials at Port Colborne, Ontario, Canada. When fire raged out of control at the eight-story Maple Leaf Mill on October 7, 1960, Buffalo Fire Commissioner Robert Howard sent the *Cotter* and crew, augmented by Engine Company 8's crew. Lacking radar and navigational equipment, the *Cotter* was escorted by a US Coast Guard cutter for the nearly two-hour long trip on Lake Erie. For more than four hours the *Cotter* pumped at its maximum 15,000 gpm, finally bringing the blaze under control.

Despite the *Cotter's* success at many waterfront fires for the following two decades, the boat fell victim to budget cuts in 1992; her firefighting status was reduced and she was used for ice-breaking only. Little did officials realize the fallout from stripping the waterfront of fireboat protection. Soaring insurance rates and the threat of jobs lost in industry left them no choice but to return the *Cotter* to full firefighting status. During the same period, city officials resisted efforts to give the *Cotter* landmark status, concerned that it might restrict their ability to alter the vessel's condition. Finally, in 1996, the *Edward M. Cotter* joined the small number of fireboats designated National Historic Landmarks. It was overhauled in 2012 and began its 112th year of service.

The *Cotter* in dry dock shows the wing bridges of the deck surrounding the pilothouse, and the vessel's deep draft, which is the portion of the boat that lies beneath the waterline. Courtesy of Richard Sikora.

Buffalo's *Edward M. Cotter* is the oldest piece of operational firefighting equipment in America. Although twice rebuilt, it retained its original form. Its immaculate appearance reflects the pride of its crew. Courtesy of Dave Organ.

Over the din of *Cotter's* engine room, engineers were alerted to orders from the pilot with this whistle system. Courtesy of Richard Sikora.

This 6,000-gpm monitor was installed on the *Cotter's* fantail during the 1953 rebuild. Its 15,000-gpm rating allows the monitor to throw a stream a distance of 600 feet. Aft of the monitor is a single tow bitt and an electrically operated capstan. Courtesy of Glenn Duda, www.allhandsworking.com.

Buffalo's *Cotter* shows off its magnificent condition at 103 years old in 2003. The two 4-inch, 6,000-gpm bow monitors are believed to be original. The projecting bow stem is part of the heavily reinforced bow for ice-breaking. From its berth on the Buffalo River, the *Cotter* has easy access to the harbor, Lake Erie, and the Niagara River. Courtesy of Roger LeLievre.

Detroit

The city of Detroit, Michigan, was founded early in American history, mainly because of its accessibility from the Atlantic Ocean. Detroit's population had swelled to nearly a quarter of a million people by the end of the nineteenth century. A paid fire department had been in place since 1860, yet it took a petition urging fireboat protection signed by owners of forty-two riverfront businesses before the fire commissioner acted in 1892 to purchase a fireboat.

The first of four large fireboats that would serve Detroit during the next 83 years was the *Detroiter.* Craig Shipbuilding Company of Toledo laid the wooden keel of hull number 57 on 11 October 1892, and it was launched on 15 May 1893. The vessel was 115 feet long, with a 25-foot beam and 9-foot draft. Two engines and boilers powered two sets of double-acting pumps enabling an output exceeding 5,000 gpm. Designated Steam Fire Engine Company No. 16, the *Detroiter* was assigned a crew of 12 and berthed alongside a new firehouse at the foot of Fifth Street. Having built an extensive high pressure pipeline that stretched from the waterfront inland to the high-value district, Detroit followed the practice of sending a hose wagon on all alarms answered by the *Detroiter.* The 1,000 feet of three-inch diameter hose carried on the horse-drawn hose wagon, along with 5,000 feet of large diameter hose on pivoting hose reels of *Detroiter's* afterdeck, proved sufficient at major fires. During its first year of service, the *Detroiter* proved worthy, having confined and knocked down major fires that otherwise would have spread along Detroit's crowded waterfront.

The potential for conflagration along the city's waterways was justification for a second fireboat. Named after the first fire chief, the *James Battle* was launched at the Detroit Shipbuilding Company at Wyandotte on 13 October 1900. With its hull and deckhouse of steel construction, the *Battle* measured 122 feet, having a 25-foot beam, and drafting 11 feet. Its two pumps, which came from the shops of Thomas Manning Jr. Co. of Cleveland, each delivered 2,900 gpm at 160 psi. The *Battle* went in service on 15 January 1901, sharing not only the *Detroiter's* quarters, but the designation "Steam Fire Engine Company No. 16."

The partnership was short-lived due to premature deterioration of *Detroiter's* wood hull. It was retired in 1902 and taken to Jenks Shipbuilding at Port Huron, Michigan, where her pumps and other machinery were removed, overhauled, and installed in Jenks' steel hull number 21. Steam engines built by Cowles Engineering Company of Brooklyn drove the pumps to supply 6,500 gpm. The new vessel steamed to

Detroit's fireboat *James Battle* slides the ways at Wyandotte, Michigan, on October 13, 1900. The all-steel, 122-foot vessel was Detroit's second fireboat. Author's Collection.

Underway at top speed, the *Battle* shows the dramatic difference in appearance after outfitting for service. The number of funnels corresponded to the number of steam boilers. Elongated pilothouse windows offered all-around visibility for the pilot and officer. Courtesy of Detroit Historical Society.

Built in 1893, Detroit's first fireboat was the *Detroiter*, which underwent complete rebuild after a decade of service. After refurbishment to a facsimile of the *Battle*, it was renamed *James R. Elliott.* Its undercut "spooned" bow was designed for ice-breaking. Courtesy of Detroit Historical Society.

The Detroit Fire Department fireboat *James Battle* works at the burning of the Belle Isle Bridge on April 27, 1915. The bridge spans the Detroit River between Detroit and Belle Isle. It was rebuilt in 1923 and renamed the Douglas MacArthur Bridge in 1942. Courtesy of Detroit Historical Society.

Detroit late in 1902 and began service at the firehouse built near the foot of McDougall Avenue to accommodate newly organized Steam Fire Engine Company No. 25. The fireboat retained the name *Detroiter* when it began service as Engine Co. 25 in February 1903, but was soon renamed *James R. Elliot* in honor of Detroit's second fire chief.

Besides fighting blazes along Detroit's miles of waterfront, the *Elliot* and *Battle* periodically worked at major fires of other cities, being no strangers to neighboring Windsor, Ontario, Canada, across the Detroit River.

If anyone could find 347 tons of floating steel endearing, it was Detroit's hardy firefighters and the citizens of Motor City when the fireboat *John Kendall* arrived. With the Detroit River ranked as the nation's busiest inland waterway, and with rapid growth along Detroit's twenty-two miles of waterfront, fire department officials sought additional fire protection and a replacement for the *Elliot*. Plans drawn in 1929 by Dean Herbert C. Sadler, professor of naval architecture at the University of Michigan, profiled one of the largest, most powerful fireboats in the nation. Few could match its power, and at 135 feet, only New Orleans' *Deluge* was longer. Named after Detroit's third fire chief, the *John Kendall* began life as hull number 182 in the Toledo Shipbuilding yard at Toledo, Ohio. Tailored for duty on the Detroit River, the *Kendall* was unique in a number of ways. The expectation that the vessel would work in severe ice conditions during winter months, along with a budget appropriation limit, influenced the decision to power the fireboat with steam engines, despite diesel power's rising popularity. Another departure from usual practice in big fireboats was fixing *Kendall's* pumps at 200 psi to avoid the delay of compounding two pumps from 150 psi at reduced output to 300 psi. This enabled *Kendall's* crew to quickly connect with the city's high-pressure mains designed for 300 psi.

No longer recognizable as the fireboat *James Battle*, it became the *Normac* after conversion for passenger, car ferry, and freight service in Canada's Georgian Bay. In 1969, at seventy-six years old, it embarked on a third career as a floating restaurant at Toronto. Courtesy of Detroit Historical Society.

The *John Kendall* was the longest lived and most revered of Detroit's fireboats. During the era of battleship-class fireboats, it was one of America's largest and most powerful. Although steam-powered, even at full power as seen here discharging fifteen streams, it expelled no smoke from its stacks since powerplants discharged into a condenser. The stack of a steamship in the background provides dramatic comparison. Author's Collection.

A portion of the *Kendall's* crew poses with their fireboat in September 1963 when Coast Guard inspectors declared the vessel unseaworthy. Morgan L. Sutherland, center, who was master of the boat for sixteen years, stated, "This boat is my pride and joy and I don't want anyone saying anything bad about her." At left is crewman Gryseels of the Detroit Fire Department, and marine engineer Harold Richards. The *Kendall's* crew of eight comprised a master, engineer, oiler, two boiler operators, and three firefighters who served as deckhands. Courtesy of Curt Elie Collection.

Two crewmen of Detroit's Fireboat 16, the *John Kendall*, secure a foundering pleasure craft struck by a freighter on the Detroit River in July 1955. Twin funnels correspond to the *Kendall's* two Babcock and Wilcox boilers. Long tubes on the aft side of stacks are relief valve stacks, made high to prevent scalding nearby crew when steam was released. Two 14-foot lifeboats are mounted between the stacks. Courtesy of Curt Elie Collection.

Toledo Shipbuilding built the *Kendall's* 950 hp steam powerplant, which turned a 9-foot, four-blade screw, giving a speed of 16 mph. A minimum of 11,000 gpm was achieved with two large and two smaller four-stage centrifugal pumps, all connected to steam turbines. During *Kendall's* acceptance tests, these pumps put out nearly 14,000 gpm, with the capability of 16,000 gpm. In a display of tremendous power, during the first official trip on the Detroit River in February 1930, with numerous officials aboard, the *Kendall's* ice-crushing bow—designed to break 17 inches of ice—easily broke through foot-thick ice. Then it was time for a novel display of *Kendall's* real power. After being backed at full speed onto the ice, the boat was stopped, a 4½-inch tip was threaded onto the monitor in the aft tower, the pumps were brought to full power, and the back-thrust from the stream was sufficient to propel the boat up to 4 mph with full steering control.

Beside the tower monitor, which was 30 feet above the waterline, pumps supplied four other 6,500-gpm Morse Invincible monitors; two amidships atop the deckhouse, and two above the pilothouse on a "winged" bridge platform. Sixteen 3-inch discharge outlets were installed on the deckhouse.

Adding to the *Kendall's* uniqueness was the use of chrome plating for monitor nozzles, railings, and numerous other fittings. The first radio equipment used by the fire department was installed in the *Kendall* during its first few months of service.

The *John Kendall* and crew of nine were berthed at the foot of 24th Street below the famed Ambassador Bridge, replacing the *Battle* as Engine Co. 16. The *Battle* then replaced the twenty-eight-year-old *Elliot,* which, although retired, had a long career ahead of it. It was purchased by Owen Sound Transportation Co., which completely rebuilt it to a dieselized vessel named *Normac.* For thirty-two years, the vessel safely carried passengers, cars, and freight between Ontario's Owen Sound and Sault Ste. Marie. In 1969 the old fireboat began another life as a floating restaurant on the Toronto waterfront.

With diesel emerging as the new source of marine power, during overhaul in 1936 the *Battle* was equipped with Cummins diesel pumps, while retaining steam for propulsion. Since the *Kendall* was deemed sufficient to go it alone, the *Battle* was decommissioned in 1940 and Engine Co. No. 25 disbanded. In May 1941, the *Battle* entered private ownership in Canadian waters. After wartime service at Nova Scotia, it languished for more than a decade in dry dock, finally undergoing rebuild as a 1,000 hp tug with firefighting capability. While with McAllister Towing from 1959 to 1992, the *Battle* was made available to the Montreal Fire Department on request.

The *Kendall,* meanwhile, was earning its reputation. In 1932, its crew put 50 million gallons of water on the Detroit Sulphite fire, and in 1941 fought flames for ninety-six hours when the gasoline tanker *Transiter* exploded on the Rouge River. Despite the *Kendall's* stellar performance at numerous blazes, including the massive Michigan Central Freight blaze in June 1966, rumblings in City Hall foretold the end of its firefighting career. Talk centered on a smaller, faster, more efficient replacement with a much shallower draft to get into places denied the *Kendall* due to its 11-foot, 6-inch draft. Although a small pump-and-monitor-equipped jetboat joined the Detroit Fire Department in 1970, insurance requirements called for a fireboat capable of pumping 6,000 gpm. Budget cuts forced the *Kendall* out of service in January 1972, although it remained available for emergencies. More financial woes brought an end to the beloved *Kendall's* forty-six years as a valiant firefighter, and it was sold to Michigan's Panoceanic Engineering Corp. in 1977.

The first black Detroit firefighter to be killed in the line of duty inspired the name of the department's new fireboat, the *Curtis Randolph.* The all-aluminum, 78-foot fireboat was built by Peterson Builders of Sturgeon Bay, Wisconsin, and placed in service in November 1979. Continuing the designation Engine Co. No. 16, the *Randolph* was stationed nearest the heaviest concentration of Detroit's waterfront industry, southwest of the city's center. While in service from May through November, the vessel is crewed by a civilian pilot and deck hand; for a working fire, a company of firefighters is picked up en route.

A pair of 12-cylinder GM diesel engines provide a speed of 23 mph, while two 425 hp Detroit diesels each drive a 3,200-gpm Patterson pump; driving engines can also be engaged to power pumps. In addition, a 2,000-gpm Worthington pump generates

Fresh from the shipyard at Sturgeon Bay, Wisconsin, the Detroit Fire Department's 78-foot *Curtis Randolph* awaits christening at the Grand Boulevard dock on May 9, 1979. Its lime green livery was later changed to white over red superstructure and red hull. Courtesy of Curt Elie Collection.

For marina protection, this 21-foot jetboat began service on October 9, 1970. It was named *Louisa St. Clair* after one of the patriotic Daughters of the Revolution. A Detroit chapter bore her name, as did the city's Lake St. Clair and the St. Clair River. Courtesy of Curt Elie Collection.

Among a long line of boat tenders used by the Detroit Fire Department was a pair of 1922 Ahrens Fox Model J-13 High Pressure Hose Cars. Designated High Pressure No. 1, this was used with the high-pressure hydrant system in downtown Detroit. Five-inch-diameter hose was laid from the fireboat to the unit's monitors or for use as hydrants. Author's Collection.

Two former Great Lakes fireboats, Detroit's *James Battle* and Chicago's *Graeme Stewart* in the livery of Canada's McAllister Towing, circa 1960. The *Battle* underwent major rebuild in 1955. The *Stewart* was scrapped in 1970. Courtesy of Curt Elie Collection.

During its first year of service, Detroit's *Randolph* demonstrates full pumping power. The 50-foot Telesqurt incorporated a ladder for rescue. Courtesy of Curt Elie Collection.

Detroit's boat tenders mounted two fireboat-size Morse monitors rated at 3,000 gpm each. Boat Tender 3 is seen here in 1971. The Detroit Fire Department also used a 1913 Ahrens Fox ladder truck tractor converted to a third boat tender that remained in service until 1952. Courtesy of Chuck Madderom.

Styles and paint schemes changed with modernization of Detroit's boat tenders, but the practice of mounting two fireboat-size monitors was retained through all models. Boat Tender 1 was shop-built using a 1973 Chevy truck. Courtesy of John A. Calderone.

400 psi, giving the *Randolph* a combined flow of 11,600 gpm at 150 psi. Four 2,000-gpm monitors are complemented by a 1,000-gpm monitor atop a 50-foot Telesqurt boom that can also deliver foam from a 200-gallon supply of concentrate.

Despite the *Randolph's* efficiency and minimal cost, Detroit's perpetual financial problems caused it to be out of service from 1983 to 1985, and again in 2012.

Having recognized from early on the value of boat tenders used in conjunction with fireboats, Detroit firefighters took full advantage of the unique vehicles to haul large-diameter hose and mount heavy stream appliances. Early horse-drawn hose wagons led to conversion of a 1913 Ahrens-Fox ladder truck tractor into a tender, followed by a pair of 1922 Ahrens-Fox boat tenders, which saw half a century of service before retirement in 1972; the 1913 unit remained in service until 1952. A third unit built locally in 1950 also served until the early 1970s, when it was replaced by a boat tender built in the department shop. Standard procedure had boat tenders responding on all waterfront alarms to which fireboats were sent.

Detroit's Boat Tender No. 1 during the 1960s was this 1950 GMC featuring body work by General-Detroit Corp. Courtesy of Chuck Madderom.

This LARC V went to the Port Clinton, Ohio, fire department in 1981, serving as a fireboat until the early 1990s. In modern times, the department operates a 23-foot, 115 hp Brigg boat, which is used mainly for EMS on Lake Erie and Sandusky Bay. Courtesy of Kent Johnson, Port Clinton Fire Department.

As a cost-saving measure, the Huron, Ohio, fire department partnered with the Sandusky Fire Department to design identical hulls for their fire-rescue boats. Huron's 30-foot Argus was built by Seattle's Northwind Marine and delivered in 2005. The all-aluminum vessel is powered by twin Evinrude 225 hp outboards. Courtesy of Capt. Kurt Schafer, Huron Fire Department.

To cover large areas of Lake Erie and Presque Isle Bay, the West Lake Fire Department of Erie County, Pennsylvania, operates this unique Husky Airboat-Husky Nattig. A Chevy V-8 engine turns an 80-inch-diameter prop to power the 23-footer. Courtesy West Lake Fire Department.

Toledo, Ohio, fire-rescue's 28-foot SeaArk Protector features a 1,000-gpm bow monitor. The all-aluminum Marine 1 is powered by twin 225 hp outboards for speeds up to 42 mph. SeaArk Marine's Protector series ranges from 26 feet to 50 feet long and up to 4,000 gpm. Author's Collection.

As a cooperative police-fire effort, Ludington, Michigan, uses this 26-foot 1,600-gpm Harbor Guard Firehawk. Twin 225 hp Mercury outboards propel Harbor One up to 60 mph. A 300 hp Mercruiser inboard engine powers jet drive and monitor pump. Eight-foot wide dive doors on both sides of the vessel fold down and float. Courtesy of Steve Smedley.

This Firestorm 36 named *Last Chance* of the Clayton, New York, fire department features a boarding ladder for divers that swivels to an upright position for stowage. Steps in three locations lead to a surround dive platform. Courtesy of MetalCraft Marine.

CHAPTER 3: NEW ENGLAND

Boston

New England was, as a whole, the most industrialized region of the US, accounting for nearly thirty percent of all manufacturing in the country by the mid-nineteenth century. Since five of the six states that comprise New England—Maine, Rhode Island, Connecticut, Massachusetts, and New Hampshire—border the Atlantic Ocean, fireboats have long been an important commodity in the region. Although the Boston Fire Department, which is the largest municipal fire department in New England, traces its origin back to 1678, fireboat service did not begin until 1873. Construction of Boston's first fireboat began in 1872, but it was several weeks from completion when, on November 9, 1872, a fire that started in the mercantile district swept the waterfront. Eighteen hours later, half of the city's waterfront was a smoldering ruin; more than 770 structures had been destroyed and fifteen firefighters had been killed.

Boston's surprisingly rapid growth during this period of industrialization triggered a steady course of fireboat purchases and assignment rotation. Its first fireboat was a benchmark in history as America's first purpose-built steam fireboat. Designed by Boston Fire Chief John S. Damrell and named *William M. Flanders* after a city council member, the fireboat arrived from Atlantic Iron Works of East Boston in December 1872, and went in service on January 1, 1873, at Central Wharf at downtown's Atlantic Avenue. Displacing 55 tons, the iron-hulled fireboat was 75

This 1923 view of Boston's Engine 44, the all-wood *Angus J. McDonald*, shows its vertical bow stem. Immediately aft of the stem post is an H-bitt for towing. Typical of large fireboats of the period were ornate trailboards and oak fenders resting on rub rails. Courtesy of Leslie Jones, Boston Public Library.

An extremely tall funnel supported by guy wires, such as that seen on the *Angus J. McDonald*, usually indicated a boiler that operated with a natural draft. During the 1920s, small powerboats were added to Boston's fireboat fleet. Courtesy of Leslie Jones, Boston Public Library.

feet long, with a 15-foot beam and a 7-foot draft. Four Amoskeag pumps delivered 2,500 gpm. On January 17, 1884, the *Flanders* took on the dual designation Engine Co. 31, serving just sixteen years before being placed on reserve status. Her replacement was the *John M. Brooks,* which had twice the gpm output of the *Flanders* and measured twice the length; it featured a 24-foot beam with a 9-foot draft. The *Brooks* went in service as Engine Co. 31 on July 1, 1889, only to be deactivated on September 1, 1895, with the arrival of the *Angus J. McDonald,* designated Engine Co. 44, a 6,500-gpm vessel measuring 110 feet.

Next was the *John P. Dowd,* weighing in at 179 gross tons and stretching 113 feet. Named after a deceased firefighter, the 6,000-gpm *Dowd* came from the yard of East Boston's Bertelsen & Peterson Engineering Company in 1909. In view of the rapid aging of the *Brooks* and the *Flanders,* naval architect William T. Keough and Boston Fire Department shop superintendent Eugene M. Byington paid careful attention to the *Dowd's* design to ensure maximum service life. They studied fireboats of other cities and characteristics of Boston's waterfront. Limited water depth at high-hazard wharfs required light draft. Heavy winter ice and use in saltwater called for the strength of a wooden hull versus plated steel. The entire white oak hull below the waterline was sheathed with copper, with yellow pine used for hull framing. Deck planks were of Oregon fir and the superstructure was built of steel, with a wooden pilothouse. Compared to fireboats of the Great Lakes, where feed water was pumped directly into boilers, those in coastal regions required more complicated machinery with condensing apparatus. Two compound steam engines, each with a Scotch boiler 11 feet in diameter and 11 feet long, powered two vertical, tandem duplex pumps to produce 6,000 gpm at 175 psi. All exhaust steam was condensed and pumped back into the boilers.

Unique among the boat's five monitors was one attached to the funnel for use as a water tower for reach over buildings and high lumber piles. Hand wheels allowed a crewman atop the pilothouse to operate the high nozzle. To supply land companies, the boat was outfitted with a dozen discharge outlets and 3,000 feet of 3½-inch hose.

On October 14, when the *Dowd* began service as newly organized Engine Co. 47, the Bertelsen & Peterson yard already had made preparations for laying the keel of the *Thomas A. Ring,* an 84-footer rated at 3,000 gpm. Innovations from lessons learned building the *Dowd* were carried over into the *Ring.* Earmarked for berthing at East Boston, the *Ring* arrived in March 1911 and began service as reactivated Engine Co. 31.

Boston's Engine 31, the *Thomas A. Ring*, at work in 1925. Author's Collection.

In those days of wooden waterfronts and few safety codes, fireboat crewmen accepted the hazards of their profession. But none could have foretold the disaster that befell the crew of the *Thomas A. Ring* on an unusually warm day in January 1919. In what is billed as the strangest disaster in Boston history, a five-story tank filled with 2.3 million gallons of molasses exploded, sending a tidal wave of the thick goo through the streets. Buildings were torn from

An ample supply of oil is fed to the boilers of the *Matthew J. Boyle* to power its pump engines when it was new in 1931. The all-steel, 12,000-gpm *Boyle* was the largest fireboat ever to ply Boston waters. Courtesy of Leslie Jones, Boston Public Library.

Fire Launches were introduced to Boston's fireboat fleet during the 1920s. The 16½-foot wooden boats were equipped with a small pump for fires in shallow areas and under piers. This fire launch in 1928 is manned by two crewmen of the *Thomas A. Ring*. Courtesy of Leslie Jones, Boston Public Library.

In this 1930 display of Boston fireboats are the *Thomas A. Ring*, at center, the *John P. Dowd*, at left, and the *Angus J. McDonald*. Although a single funnel usually indicated a single boiler, dual steam powerplants of some Boston fireboats breeched into a common stack. Courtesy of Leslie Jones, Boston Public Library.

their foundations and the vacuum created by the wave pulled a train from the tracks. The massive wave shoved nearby Engine Co. 31's station 10 feet toward the water and collapsed interior floors, trapping five of the fireboat's six crewmen. Marine Engineer George Layhe died in the collapse, joining the list of twenty others killed on that fateful day of the "Great Boston Molasses Disaster."

So confident was the crew of Engine Co. 47 with their new boat's capability that they attacked a monstrous warehouse blaze on the leeward side with all five monitors, confining the fire that threatened to destroy the entire waterfront. Satisfied that their waterfront had ample fire protection, Boston officials would not consider fireboat acquisitions for the next two decades. Punctuating the fireboats' effectiveness was the replacement in 1921 of fireboat-supplied saltwater mains with high-pressure pumping stations. The decade also saw the introduction of 16½-foot launches, each equipped with a 300-gpm pump for fires in shallow water and under piers. These "fire launches" would be placed in service with Boston fireboats as standard equipment until the 1950s.

Built into the Boyle's stout frame were tow bitts on its after-deck and bulwarks. It was not uncommon for large fireboats to tow disabled or burning vessels. One of the *Boyle*'s seven monitors was atop a sturdy tower. Courtesy of Leslie Jones, Boston Public Library.

After twenty years of fighting flame along the waterfront, the time had come to update the fireboat fleet. A search by fire department officials led them to the George Lawley Shipyard of Quincy, Massachusetts, which on May 23, 1931, launched the *Matthew J. Boyle.* At 125 feet long and rated at 12,000-gpm, this was the largest and most powerful fireboat ever to ply Boston waters. Power for the 339-ton vessel came from oil-fired steam engines. Built entirely of steel, she sported five monitors, one of which was mounted atop a sturdy steel tower on her deck house. The *Boyle* went in service on December 8 as Engine Co. 44 at India Wharf, replacing the *McDonald,* which became Engine Co. 31 until its retirement in 1947. The *Ring* then went into reserve until 1941, when it became Engine Co. 47, and was retired in 1943.

On March 10, 1937, Boston fireboats battled fire that consumed nitrate-laden freighter *Laila* at Mystic Docks. Fireboat crews often worked for hours dangerously close to volatile cargoes. Author's Collection.

Fireboats regularly supported dive operations. A davit of the *Matthew J. Boyle* is used to lower a diver in 1937. A launch on the top deck is marked 1 Tender. Radio antennas are visible running from the pilothouse to stand-offs on the funnel and monitor tower. Courtesy of Leslie Jones, Boston Public Library.

In marine tradition, early Boston fireboats wore elaborately carved trailboards bearing the vessel's name and incorporating the hawspipe. Unusual on the *Thomas A. Ring* was a monitor affixed to the funnel, with controls atop the pilothouse. Numbered discharge outlets were at the base of the deckhouse. Courtesy of Leslie Jones, Boston Public Library.

Some land apparatus that often worked in conjunction with fireboats proved as durable as the boats. This never-say-die high-pressure hose and booster wagon of Boston's Engine 26 continued pumping, even after near total destruction by a falling wall. Even more astounding was that the 1947 Mack was driven to the shop under its own power. Author's Collection.

Like many cities, Boston took advantage of the availability of low-cost military surplus after WWII by purchasing two US Navy coastal minesweepers built in 1941; USS *Marabout* (AMc-50), built by Herreshoff Manufacturing, and USS *Bulwark (AMc-68),* built by Hodgen Bros. & Goudy & Stevens of East Boothbay, Maine. Both diesel-powered vessels displaced 195 tons and measured 97 feet, with a 22-foot beam and 9-foot draft. After Boston purchased the pair in late 1946, they went to General Shipbuilding Co. of Boston, which converted them to 6,000-gpm fireboats. The *Marabout* was named the *Joseph J. Luna* (Engine Co. 47), to honor a Boston firefighter killed in action during WWII, and the *Bulwark* became the *James F. McTighe* (Engine Co. 31).

Boston fireboat crews previously had been housed in stations; however, since a station remained only for the crew of the *Boyle,* the *Luna* and *McTighe* were built to house their crews. Both vessels were berthed at Battery Wharf at the North End. The pair teamed with the aging *Boyle* until it was retired in 1952 and Engine Co. 44 deactivated. After a quarter century of yeoman service, the days of the sister fireboats were numbered when plans were laid for their replacement. With a new fireboat slated for arrival in 1972, the *McTighe* was sold, and the *Luna* went into reserve status as Engine Co. 31 until 1976, when it too was sold.

Converted from the navy coastal minesweeper USS *Marabout*, Boston's fireboat Engine 47, the *Joseph J. Luna*, lies alongside a huge shed used to store coal for the Lincoln Power Station at Boston's North End. The 97-foot *Luna* began service in 1948 and went in reserve when replaced in 1972. Courtesy of Fire Apparatus Journal Collection.

Moored at Boston's Lincoln Coal Pocket, the *James F. McTighe* displays its early configuration with monitor tower that also served as a mast incorporating a floodlight, tow lights, and a salvage boom. Conversion from the minesweeper USS *Bulwark* to fireboat by the Boston shops included a raised pilothouse with afterhouse. Courtesy of Fire Apparatus Journal Collection.

This 1971 view of the *McTighe* shows it reconditioned with a new bow fender and radar atop the pilothouse. When replaced by the *Firefighter* the following year, the *McTighe* went into commercial use. Author's Collection.

The 76-foot *Firefighter* replaced Boston's pair of surplus minesweepers-turned-fireboat. The *Firefighter's* five original Stang monitors were soon replaced in favor of those removed from outdated hose wagons. Two discharge manifolds are mounted forward of the pilothouse. Courtesy of Mike Legeros.

The *Firefighter* displays its full power in 1992 with streams from five topside monitors and two under-pier nozzles. Briefly mounted next to the pilothouse was an articulating boom with nozzle, which soon succumbed to saltwater corrosion. Courtesy of Bill Noonan.

Taking their place was an all-steel, 6,000-gpm fireboat measuring 76 feet long, with a beam of 19 feet. Drafting only 5 feet, 6 inches, and 19 feet in height above its waterline, it was able to pass under many of Boston's bridges, except at high tide. Named simply *Firefighter,* the vessel weighed 125 gross tons. Two of its four 525 hp Detroit diesel engines drove two 3,000-gpm pumps, while two were dedicated to propulsion, allowing the crew to pump at full volume while maneuvering and attaining a top speed of 14 mph. Besides five monitors topside, which came from outdated hose wagons, the *Firefighter* featured two under-pier bow monitors. Also aboard were 500 gallons of foam concentrate and a 13-foot Boston Whaler with an outboard motor, plus a small aluminum boat. During construction in 1971, a 55-foot TeleSqurt boom was mounted immediately aft of the pilothouse; however, over time, salt air and water so badly corroded the system that it was removed.

The mooring lines of the *Luna* were cast off for the last time in 1976 when the 45-foot *St. Florian* arrived from the Damariscotta Boat Company, Rockland, Maine. Named after the patron saint of firefighters, the 3,000-gpm steel vessel featured a 16-foot beam and drafted only 4 feet. A pair of 318 hp Detroit diesel engines simultaneously drove two Niagara pumps and propelled the boat up to 17 mph.

With the *St. Florian's* arrival, the historic designations Engine Co. 31 and Engine Co. 47 passed into history, and the *Firefighter* and *St. Florian* became known as Marine Units 1 and 2, respectively. Both boats responded to all waterfront alarms. Budget woes caught up with the Boston Fire Department during the 1980s and twenty companies, along with the crew of the *St. Florian,* were eliminated. The vessel went on reserve status and in 1992 was replaced by the *St. Florian II,* a smaller boat at 30 feet and less powerful, rated at 1,000 gpm. Twin 200 hp Volvo diesel engines gave the boat a top speed of 35 mph.

As modernization dictated even smaller, less powerful vessels, in 2006 the department took delivery of a 27-foot, 700-gpm Boston Whaler Challenger, which was named the *Norman McKnight.* In 2008, the *St. Florian II* was donated to the fire department of New Bedford, Massachusetts. The Boston Fire Department later placed in service a 27-foot Boston Whaler Vigilant with a 600 hp engine and 200-gpm Darley pump.

In 2011, after nearly forty years of service, the *Firefighter* was retired, with its replacement from the long line of popular fireboats built by Canada's MetalCraft Marine. Named the *John S. Damrell* after Boston's fire chief during the great fire of 1872, the all-aluminum "Firestorm 70" features a 23-foot beam and 3-foot draft. Four 825 hp Iveco engines drive four Hamilton HJ364 waterjets, giving the boat a top speed of 46 mph. Four Hale pumps provide 12,000 gpm for five monitors, including a remote-controlled Stang and four Elkharts, two of which are remote-controlled.

Fireboats of the Boston Fire Department share responsibility with those of the Massachusetts Port Authority, called Massport, which serves Boston Harbor and specifically, New England's largest and busiest airport, Logan International. The airport is essentially in the middle of Boston Harbor, with all approaches and departures made over the harbor. Massport's fireboat, the *Howard W. Fitzpatrick,* was a twin to the Boston Fire Department's *Firefighter,* both having been built side by side in 1971 at Grafton Boat Works, Grafton, Illinois. They traveled together on their delivery voyage, down the Mississippi, across the Gulf, then up the coast to Boston. Like the *Firefighter,* modernization caught up with the *Fitzpatrick* in 2011, when it was replaced in November by the 79-foot *American United,* whose name was derived from the airlines operating the two ill-fated airliners that departed Logan International on the morning of 9/11.

Refurbishment of Massport's *Fitzpatrick* in 2007 included a new paint scheme, removal of the Telesqurt, and a section of the bulwark removed to facilitate dive operations. Hose was kept in aft compartments with roll-up doors. Courtesy of Jay Albert's Cape Ann Images.

Boston Massport's modern fireboat was named *American United* after the two ill-fated airliners that departed Logan International Airport on the morning of 9/11—American Airlines Flight 11 and United Airlines Flight 175. It is seen here in December 2012. Courtesy of John Galla.

Boston Massport's *Howard W. Fitzpatrick* is a twin to Boston Fire Department's *Firefighter.* The 133-gross-ton fireboat seen in its original configuration included a white over blue paint scheme, and a 55-foot Telesqurt mounted immediately abaft the pilothouse. Courtesy of Curt Elie.

The patriotic emblem that adorns the superstructure of *American United.* Courtesy of John Galla.

Boston Fire-Rescue's *John S. Damrell* is a 70-foot MetalCraft Marine vessel that replaced the Firefighter in 2011 after forty years of service. Courtesy of Andre Blanchard.

This fire pump-equipped twin-Vee catamaran served Massport with a crew of two. Courtesy of John Galla.

The all-aluminum vessel was built as a Ranger 2400 Model familiar to A. F. Theriault & Son of Meteghan River, Nova Scotia. Four Caterpillar C32 outboard diesel engines rated at 1,450 hp each, along with a pair of inboard C18 engines rated at 873 hp each, give the vessel a top speed of 28 mph and an output of 6,000 gpm. Its 7-foot draft hull is reinforced for operations in ice and incorporates 75 hp bow thrusters. Firefighting equipment includes a Stang 3,000-gpm remote-controlled monitor, two forward 1,500-gpm remote-controlled Stangs, and two manual-controlled 1,500-gpm monitors aft. A complete EMS suite includes thirty Switlik life rafts for 300 survivors of an aircraft crash.

Since the 1970s, Massport's emergency boat inventory has included small vessels such as Boston Whalers ranging in length from 21 feet to 27 feet. The authority's *Marine 32* is a 37-foot catamaran built by Moose Boats in 2009. The twin-vee hull boat features a 500-gpm pump for its bow monitor and it can attain a speed of 46 mph.

After Boston provided New Bedford its first fireboat, it was renamed *Robert Gonsalves,* and augmented by New Bedford's *Edward C. Spooner,* a former US Coast Guard 41-footer. New Bedford shares coverage of Buzzards Bay inner harbor with Fairhaven's 25-footer, also a former USCG vessel. Other Massachusetts fireboats include those of Dartmouth, Plymouth, Winthrop, Beverly, and Cambridge, while the Cape Cod region enjoys substantial coverage by numerous fire-rescue boats of nearly forty fire departments. A cooperative agreement in early 2014 enhanced the effectiveness of this network with formation of Buzzards Bay Marine Task Force.

Portland

Portland, Maine's, experience with devastating fires is reflected in its city seal with a phoenix rising from the ashes. Significant among the blazes in the history of Maine's largest city is the 1866 Great Fire of Portland. After it was ignited from July Fourth fireworks in a Portland Harbor boathouse, the blaze spread across the city, destroying more than 1,500 buildings. This was the largest conflagration yet seen in a US city. The catastrophic event resulted in the installation of a fire alarm signal system and augmentation of the city's water supply from Sebago Lake. Since fireboat design was rudimentary, at best, it wasn't until 1894 that Portland's fire department considered fighting fire from the water. The city purchased the 1891 ferryboat *Chebeague* from the Harpswell Steamboat Company at Casco Bay, Maine, and reconfigured it to serve as the fireboat *City of Portland*—Engine 7. A new 1,800-gpm Amoskeag extra first-size pump and two monitors were installed, and the boat was commissioned on February 19, 1895. Berthed at the Portland Pier, the vessel protected the busy waterfront during the city's growth as a major rail hub until 1930, when the time for change had arrived.

In 1931 a new fireboat assumed both the original fireboat's designation and the role of protector of Portland's waterfront. Built by Rice Brothers Corp. of East Boothbay, Maine, the 90-foot, 115-ton vessel could discharge 6,000 gpm. The vessel was the nation's second diesel-powered fireboat, the first having been built for Houston six years earlier. Houston's differed in being diesel electric, which used a diesel engine to power electric motors that turned a prop shaft.

The nation's second diesel-powered fireboat was the *City of Portland* of Portland, Maine. The first diesel fireboat was built for Houston, although it was diesel-electric. The 90-foot *City of Portland*, seen here in 1950, was built in 1931 and rated at 6,000 gpm. Courtesy of Maine Historical Society.

Portland, Maine, fire department's third fireboat named *City of Portland* was this 65-foot, all-steel, diesel-powered vessel delivered in 1959. Its original color scheme was red over black. A small Lund boat is stored aft of the deckhouse. Courtesy of John A. Calderone.

In April 1959, Portland's third *Engine 7* arrived from East Boston's General Ship and Engine Works. The boat was designed by noted naval architect John G. Alden, with the hull engineered by Gladding-Hearn Shipbuilding of Somerset, Massachusetts. At 87 gross tons, the all-steel boat was 65 feet long, with a 15-foot beam and 6-foot draft. Power was derived from two Cummins V-12 560 hp diesels, with two Cummins 160 hp auxiliary engines. Two 3,500-gpm De Laval pumps could discharge 7,000-gpm through four Invincible deck monitors.

The biggest blaze worked by Engine 7 was the 1969 Grand Trunk Railroad grain elevator fire, during which a harbor tug nudged against the fireboat so that it could maintain position.

In keeping with the national trend for providing emergency medical services, in 1993 the Portland Fire Department added a 42-foot Duffy custom boat

Portland's ultra-modern fireboat, and its fourth named *City of Portland* is this 3,500-gpm vessel, which arrived in 2009. Along the top of the deckhouse is the legend, "Serving the Islands and Waterfront Since 1894," referring to the year that fireboat service began. Author's Collection.

powered by a 540 hp Caterpillar diesel engine. Named the *Joseph C. Cavallaro,* the boat served the waterfront and Casco Bay islands. As the main link to the islands, it also carried city workers and supplies, even the dead, earning it the name The Big Red Hearse.

On September 21, 2009, the *City of Portland IV* was commissioned, replacing the fifty-year-old fireboat, which was sold the following year. Built by Nova Scotia's A. F. Theriault & Sons, the 65-foot, 3,500-gpm fireboat joined the *Cavallaro* at the Maine State Pier.

At the northernmost tip of New Hampshire's eighteen miles of coastline lies Portsmouth Naval Shipyard, the navy's oldest continuously operating shipyard. After WWII, when the facility became the center for submarine development, the Portsmouth Naval Shipyard Fire Department enlisted the services of the large, 110-foot harbor tug *Tuscola* (YTB 280), which fought fires along the New England coast until 1963. In December 2006, the Portsmouth Fire Department took delivery of a 33-foot Firestorm built by MetalCraft Marine. Fire Chief Steven Achilles adds, "The boat was purchased through a New Hampshire Homeland Security grant. It is cross-staffed by our engine and Ambulance 1 companies out of our Court Street station. It is the only municipal fireboat that runs on the Piscataqua River." At the river's mouth is the island town of New Castle, whose fire department operates Marine 2, a 250-gpm Boston Whaler.

In 2012, numerous organizations formed a cooperative agency to protect the port of Providence, Rhode Island, and the greater Narragansett Bay area. Included in the Narragansett Bay Marine Task Force are fire-rescue vessels of the Coast Guard, law enforcement, and the municipalities of Providence, East Providence, Cranston, Warwick, North Kingston, Portsmouth, and Barrington. Pooling the resources of various emergency agencies caught on in other port-saturated regions of New England. Following on the heels of the Buzzards Bay Marine Task Force activated in May 2014 were marine task forces for Cape Cod Bay and Nantucket Sound.

New Haven

New Haven sits at the midpoint of Connecticut's southern coastline of Long Island Sound. As the state's second largest city, and having a large, deep harbor, New Haven sought waterfront protection prior to WWII, finally acquiring a fireboat in 1941. Under supervision of the fire department shop, a former police harbor patrol boat was reconditioned and outfitted with an 1,100-gpm pump and two monitors. It was christened the *John W. Murphy* and began service assigned to Engine 5 on October 8, 1941. The *Murphy's* replacement arrived in 1962 from the Norfolk Shipbuilding & Drydock Company. Named the *Sally Lee* after the daughter of New Haven Mayor Richard C. Lee, the vessel featured a pair of two-stage De Laval pumps rated at 3,500 gpm each. During its first

New Haven, Connecticut's, 7,000-gpm *Sally Lee* was one of the few fireboats named after females. Similarities to the *City of Portland* designed by Boston's John G. Alden are apparent in the *Sally Lee.* Its bulwarks were built with ample freeing ports. Author's Collection.

This 36-foot Firestorm was New Haven's first fireboat in more than twenty years after the *Sally Lee* was retired. Marine 1 began cooperative fire-police service in June 2013. On its transom is the name *Nathan Hale.* Courtesy of Glenn Duda, www.allhandsworking.com.

Norwalk, Connecticut, operates this 42-footer named *Robert Bedell*, which North River Boats delivered in 2012. Its aft deck steps down to a spacious dive platform. Courtesy of Britton Crosby.

year of service, the *Lee* averaged a call per week, nearly half of them for fires. After retiring the *Lee* during the early 1990s, New Haven would not purchase another fireboat until 2013 when it took delivery of a 36-foot Firestorm from MetalCraft Marine. Branford, just eight miles east of New Haven, since January 2007 has operated *Fire 1,* a 30-foot, 1,500-gpm fireboat built by Derecktor Shipyards of nearby Bridgeport, Connecticut.

Bridgeport, despite its size and location on the sound, would not take delivery of a fireboat until 2008. The city's 28-foot, 1,250-gpm Firebrand is the only fireboat between New York City and Boston. When Norwalk, a coastal city southwest of Bridgeport, took delivery of a 42-foot, 2,400-gpm fireboat in 2012, its 24-foot *Harry E. Brower* was placed in reserve. Norwalk's *Marine Unit 238* was built by North River Boats of Roseburg, Oregon. At its dedication on November 12 it was named the *Robert Bedell,* after a deceased firefighter commended for heroism at a fire in 1986. Powered by twin 425 hp diesel engines giving it a speed of 40 mph, the *Bedell* serves as a regional asset for communities along Long Island Sound.

The Noroton Fire Department of Darien, Connecticut, operates its 27-foot Boston Whaler with a 750-gpm Darley pump to extinguish a boat fire in 2009. When placed in service in 1989, Marine Unit 34 was one of the first fireboats in waters of southwest Connecticut. Courtesy of Noroton Fire Department.

Connecticut's West Shore Volunteer Fire Department operates this unique Sealegs fire-rescue boat, complete with fire pump. The amphibious rigid inflatable features retractable, steerable wheels powered by a 24 hp motor. It is sturdy enough to rest on its hull with wheels retracted. Water speed is 40 mph, with 4 mph possible on land with wheels extended. Courtesy of Glenn Duda, www.allhandsworking.com.

Although rated at 1,740 gpm, this 33-foot Firestorm of the Portsmouth, New Hampshire, fire department is capable of pumping 2,000 gpm. The vessel is named *Sagamore* after the fire department's first steam engine. Courtesy of MetalCraft Marine.

CHAPTER 4: ATLANTIC COAST

New York

By virtue of its size and expansive waterways, New York City counts more fireboats in its historical annals than any municipality in America. New York's procession of battleship-class fireboats has long captured the essence of marine firefighting operations. Important to the early history of New York's fireboat fleet are the fireboats of neighboring Brooklyn, which, during the late nineteenth century, boasted a fire department almost as large as New York's. Before the two cities were consolidated in 1898, the Brooklyn Fire Department relied on assistance from fireboats of the New York Fire Department. In 1885 Brooklyn officials met with noted naval architect William C. Cowles, who had designed many of the large fireboats during the late nineteenth century. Their collaborative efforts resulted in a 99-foot steam-powered vessel built by South Brooklyn's Trundy & Murphy. Two Clapp & Jones pumps gave it a rating of 3,360 gpm. Named *Seth Low* after a former mayor, it was commissioned January 1, 1886, and placed in service with a crew of eleven as Brooklyn's Engine 23. "The Low" was the first in its class of twenty-five large fireboats across the US to use twin screws for maneuverability. Although of all-wood construction, its hull was sheathed with copper up to the waterline. Seven years later, the Brooklyn Fire Department took delivery of a larger and more powerful fireboat named *David A. Boody*. Palmer & Son of Noank, Connecticut, built the all-wood, 105-foot vessel, which went in service as Engine 32 on October 28, 1893. The *Boody* was powered by a coal-fed compound steam condensing engine, which also drove two 3,250-gpm pumps built by the American Fire Engine Co.

Like Brooklyn, Long Island City would be absorbed into New York City; however, prior to consolidation in 1898, Long Island City had contracted with Newtown Creek Towing Company for on-call use of their tug

With the exception of a bow monitor and aft hose reel, Brooklyn's first fireboat, the *Seth Low*, appeared a conventional tug. Massive tow bitts and the traditional eagle reinforced the tug image. The *Low* served as fireboat for both Brooklyn and New York before retirement in January 1916. Courtesy of John Landers-Beth Klein Collection.

New York's Engine Company No. 23, the fireboat *Seth Low*, shortly after commissioning in 1886, with crewmen. Moustaches for firemen were then in vogue. Author's Collection.

Protector. The aptly named tug, along with five other tugs owned by the firm, was equipped with a pump, hose, and salvage and wrecking gear. Only five years old at the time of consolidation, the 53-foot *Protector* was consigned to the New York Fire Department.

Prior to establishment of the New York Fire Department in 1865, the city relied on the police department's steam-propelled side-wheeler *Seneca*, which was equipped for fighting fires. Named after a Native American tribe of western New York, the 215-ton wooden boat was built in 1862. Assigned to the 24th Precinct New York Harbor Police, the *Seneca* and six rowboats left their North River berths daily to patrol the waterways. The large side-wheeler would continue her vigil until the arrival of New York's first purpose-built fireboat. The *Seneca* then saw only periodic use until the early 1870s and was destroyed by fire in 1881.

With the newly organized fire department, the Metropolitan Board of Fire Commissioners sought a "floating engine" and soon arranged for on-call lease of a salvage tug owned by Baxter Wrecking Co. Built in Philadelphia in 1864, the 78-foot *John Fuller* was equipped with two Amoskeag pumps. Although the boat's gpm is undetermined, records note its ability to supply thirteen streams. The *Fuller* served the department until the city purchased its first fireboat, occasionally serving as a replacement fireboat from its Manhattan berth. In the meantime, the Tweed Charter of 1870 was enacted, abolishing state control over New York City; the fire department was then officially named Fire Department of the City of New York, which explains the legend "FDNY" that remains in use.

Until the New York Fire Department purchased its first fireboat, the New York City Harbor Police *Seneca* was capable of fighting fires. The 105-foot side-wheeler's beam spanned 33 feet and drew 12 feet of water. It performed police and fire service until the early 1870s and was destroyed by fire in 1881. Courtesy of John Landers-Beth Klein Collection.

The tug *Protector* was hired by the Long Island City Fire Department when needed for firefighting. Built at Long Island City in 1894, it was owned by Newtown Creek Towing on the East River. In the background is the steamship SS *Steel Pier*, which sailed the Boston-Provincetown run. Courtesy of John Landers-Beth Klein Collection.

The *William F. Havemeyer* was the first fireboat built for New York's fire department. Its coal-fired engines powered Amoskeag pumps that delivered 6,000 gpm. Large, swiveling hose reels occupied its fore and aft decks. Courtesy of John Landers-Beth Klein Collection.

The New Yorker in its original configuration with unique heat shields along stern bulwarks. Square deckhouse windows were replaced with portholes when on-board living quarters were eliminated. The bow monitor was actually a large, hose-fed nozzle affixed to tow bitts. Author's Collection.

Recognizing the need for a pure fireboat to protect its bustling waterfront—besides bolstering its image as the largest fire department in the nation—New York city fathers ordered from River Iron Works (later John H. Dialogue & Son) of Camden, New Jersey, a 106-foot, wood-hull fireboat. The yard's hull number 183 was named *William F. Havemeyer* and began service as newly organized Engine 43 with a crew of twelve on May 12, 1875. Berthing sites were of little consequence since the city had yet to provide land quarters, allowing the vessel to be positioned where it was needed. The boat was dispatched by a telegraph connection with the fire department communications center. New York's Department of Docks operated a steam ferry named *Manhattan* that assisted the *Havemeyer* at major fires. Like most fireboats prior to the early twentieth century, the *Havemeyer* took its name from a former New York mayor. Dialogue became one of the nation's four top shipbuilders, putting to sea 310 vessels, including three fireboats, before closing its doors in 1913.

It was only a matter of time before New York purchased a second fireboat, the keel of which was laid in 1882 at the Pusey & Jones yard at Wilmington, Delaware. Named the *Zophar Mills* after the assistant engineer of New York's earlier volunteer fire department, the iron-hull, 120-foot boat went into service as Engine 51 at Pier 42, North River on April 14, 1883. Coal-fired steam engines powered two Clapp & Jones pumps to deliver 6,000 gpm. The *Zophar Mills* would serve New York for more than half a century before disposal in September 1934.

Encouraged by the success of its fireboat fleet, New York Fire Department officials and designer Charles H. Haswell teamed with local shipyard Julius Jonson to design what was then the most powerful fireboat, which would reign as the flagship of the fleet for many years. Named *The New Yorker,* the steel hull vessel combined the best of materials to steam the Hudson and East Rivers and deliver a 13,000-gpm punch. Triple-expansion steam engines from Jersey City's Brown & Miller produced 150 pounds of pressure in two McNeil & McLoughlin Scotch boilers that measured 15 feet long and 12 feet in diameter. These powered two screws, one fixed and the other swiveling (to assist the rudder in maneuvering), along with two sets of pumps, one set built by LaFrance Fire Engine Company and the other by Clapp & Jones Manufacturing. The 126-foot vessel was built mainly of iron and steel, with white oak and locust woods used for interior furnishings; the deck was constructed of 3-inch-thick white pine. *The New Yorker* featured protective systems that leave no doubt as to the amount of heat encountered at fires. Most unique

In a display of the fireboat fleet, *The New Yorker* shows off its checkerboard decks that identified its flagship status. On the trunk decks of the boats are monitors with manifolds and hose reels on swiveling tracks. Alongside *The New Yorker* are the *Willett* and the *Duane*. Author's Collection.

Engine Company No. 57 *The New Yorker* in its final configuration, with a rebuilt funnel and fewer monitors. The low deckhouse rose only 3 feet above the water. It is moored at the seawall of New York Aquarium at Battery Park. Author's Collection.

among the systems were six "traveling screens" to protect firefighters operating nozzles along the rails. Made of double sheet steel with a 1-inch air space, the shields had peep holes and cutouts for nozzles. Riding on rollers, the shields could be removed from their track for additional protection at the opposite side of the deck. Fixed spray pipes about the deck protected crewmen from heat, and removable screens with peep holes were provided for the pilothouse.

While docked, 50 pounds of steam was maintained in one boiler and the other kept ready for lighting. A horse-drawn boat tender loaded with 1,500 feet of 3¾-inch hose responded on fireboat alarms, meeting the boat at dockside to supply water to blazes well inland. Expansive living quarters aboard *The New Yorker* were eliminated during overhaul about 1901 thanks to construction of a shore station at Battery Park, Pier 1, North River. Construction of the landmark Victorian structure was completed in 1886 and occupied mainly by the police department (which used the tower as a lookout) and Department of Docks. Not until 1959 would the Department of Docks leave the site and relinquish it to the Department of Marine and Aviation for use as headquarters of the fireboat fleet. Although the fire officer and firefighters assigned to *The New Yorker* occupied the new quarters, the pilot, engineers, and stokers often lived aboard to quickly get the boat underway. These fireboat operators were civilians until 1907, when they were inducted into the fire department's uniform force.

New York's sixth fireboat at the turn of the century when it was named after then-mayor Robert A. Van Wyck. In 1902 it was re-named *William L. Strong*, the name under which it was launched. The long deck house, which comprised captain's cabin, nozzle room, lamp room, and crew quarters, was removed within its first year of service when it was decided to provide quarters ashore. Courtesy of John Landers-Beth Klein Collection.

The *Capt. Connell* was the FDNY's second large fireboat tender, which served from 1922 to 1938. The 57-foot wood-hull vessel was powered by a 150 hp coal-fired compound steam engine. Courtesy of John Landers-Beth Klein Collection.

Continuous growth along New York's waterfront meant continuous growth of the fireboat fleet, giving rise to the fire department's purchase of a sixth fireboat. The 110-foot vessel began life as hull number 342 at the yard of Dialogue's River Iron Works. Built almost entirely of steel and powered by a 460 hp coal-fired, compound steam engine, she slid the ways on November 27, 1897 as the *William L. Strong.* When Robert Anderson Van Wyck succeeded Mayor Strong in January 1898, the boat was renamed in his honor. When his term ended in 1902, the boat's name was changed back to *William L. Strong.*

Two duplex pumps built by the American Fire Engine Company supplied 7,000 gpm. Designer H. de B. Parsons kept the boat's layout simple, with the only "fancy work" being the pilothouse finished in cherry wood and a wooden deck; iron bulkheads were painted to resemble ash wood. To afford the crew maximum room and ventilation, a long deck house was provided, only to be removed six months later since land stations were being built. The *Strong* began service on June 1, 1898, as Engine 66 at Grand Street, East River, logging fifty years of service when it was decommissioned in May 1948.

From early on, officials of the New York Fire Department recognized the value of fireboat tenders, which were small boats that assisted battleship-class fireboats. Often they were used as transports or as command posts, and some had limited pumping ability. Records show the first tender going in service in 1893, with sixteen tenders used throughout the city's boroughs for the next forty years. Thereafter, at least one primary tender was kept operational. The first of the large tenders was the *Velox,* which was built in Bayonne, New Jersey, and acquired from Tidewater Oil Co. in 1907. The 68-foot wooden vessel was steam-powered and served the department until 1922. Her replacement was the *Capt. Connell,* a 57-footer built by Camden Shipbuilding in 1905. It would serve until 1938 when replaced by the *Smoke,* which could pump 125 gpm.

Just months before the *Havemeyer* was retired in 1901, fireboat crews responded to one of the most disastrous loss-of-life fires in New York waters. On June 30, 1900, New York Fire Department units, including *The New Yorker* and *Zophar Mills,* arrived at a box alarm to see disaster playing out across the North (Hudson) River at Hoboken Docks. Numerous vessels and portions of the waterfront were ablaze, with some of the larger ships drifting toward New York shoreline. Chief of the New York Fire Department, Edward F. Croker, who was aboard *The New Yorker,* called for the fireboat *Strong* to aid in rescue and firefighting. *The New Yorker's* crew, using their deck heat shields, pulled alongside the blazing ocean liner *Bremen* to pull twenty-eight people through portholes. Many died because other portholes were too small or red hot. The death toll of nearly 400 would have been higher had it not been for the heroics of fireboat crews. Disasters such as the Hoboken Docks fire, which began in cotton bales and spread to barrels of whiskey, was the impetus for enlarging the fireboat fleet.

Next on the drawing board was Parsons' design for a steam-powered, 117-foot steel vessel. Named

The 7,000-gpm *Abram S. Hewitt* lies alongside a cargo barge at Pier 26 Battery Park in December 1957. Engine Company No. 77 was organized when the Hewitt began service at West 13th Street in 1903. During its career, its twin funnel was changed to a single stack and its deckhouse was altered. Colors were white over black with a buff funnel, which replaced red over black. Author's Collection.

for New York City Mayor Abram Stevens Hewitt, the 7,000-gpm boat went in service as Engine 77 on October 25, 1903. The *Hewitt's* sister ship, *George B. McClellan,* was under construction at the Camden yard when the city suffered another marine disaster, the worst in lives lost. On June 15, 1904, the excursion steamer *General Slocum* burned, killing more than 1,000 of the 1,400 passengers aboard. Fireboat crews rescued many, but their primary duty that dark day was recovering bodies. The *McClellan,* which was named after a Civil War general who later became governor of New Jersey, began service as Engine 78 on October 1, 1904. The *Abram S. Hewitt,* Engine 77, then relocated to Brooklyn.

In June 1905, Chief Croker had the seven marine engine companies organized as the 19th Battalion under the 1st Division. Having a keen interest in waterfront firefighting, Croker was sold on the concept of a sizable fireboat fleet to work in conjunction with land companies. Long-range plans already were in the works to expand the fleet to ten fireboats, which became reality the following year when three keels were laid at T. S. Marvel's shipyard at Newburgh, New York. The first two of the trio, hull numbers 185 and 186, were identical, weighing 326 gross tons and measuring 132 feet, with a 28-foot beam and drafting 10 feet. They were the *Thomas Willett* and *James Duane,* respectively. Chief Croker's dedication and passion for improving the fire service was reflected in their design, which included steam turbine driven centrifugal pumps. Among Croker's recommendations for the new boats was his statement:

> CENTRIFUGAL PUMPS, WHEN TURBINE DRIVEN, WILL DO THE SAME WORK WITH HALF THE STEAM REQUIRED BY RECIPROCATING UNITS. I THEREFORE BELIEVE THAT CENTRIFUGAL PUMPS HAVING A CAPACITY OF AT LEAST 4,000 GPM BE INSTALLED AND THAT THEY BE DRIVEN BY TURBINE ENGINES (WWW.FIREMUSEUMNETWORK.ORG).

Construction of the *James Duane*, and its sister fireboats, *Willett* and *Lawrence*, signaled the end of long deck houses, permitting more deck space. The *Duane*, Engine Company No. 85, is seen here at Pier 75, North River. Author's Collection.

This 1913 view of the *James Duane* shows Chief Croker's influence in fireboat design. His recommendations included open decks, small but high pilothouses, fixed steel skeleton monitor towers, and numerous manifold outlets. The *Duane*'s pilot, Richard Gorman, poses in the tower. Courtesy of John Landers-Beth Klein Collection.

Accordingly, the Willett and Duane featured two 600 hp steam turbines driving two-stage centrifugal pumps that tested at more than 9,600 gpm at 150 psi. For high-pressure streams to bore into large blazes, pumps could generate 4,500 gpm at 300 psi. The added benefits of steam turbines were less vibration, which strained structural integrity, and reduced maintenance, thanks to fewer moving parts.

Croker also pressed for more and larger monitors, along with additional manifold outlets for shore hose lines and a monitor in a fixed elevated steel tower. And he insisted on relatively bare, spacious decks and a small, but high pilothouse for optimum visibility.

The third fireboat of the order, Marvel's hull number 190, which was named *Cornelius W. Lawrence,* was of more conventional design since it was earmarked for duty in the Harlem River. Hence, it was smaller

The tender *Smoke*, lying alongside the fireboat *Lawrence*, served New York's fireboat fleet for seventeen years beginning in 1938. A Sterling gasoline engine provided propulsion and powered a 125-gpm pump. The *Smoke* measured 53 feet long, with a beam of 7 feet and 3½-foot draft. Courtesy of John Landers-Beth Klein Collection.

at 104 feet long and weighed 172 gross tons. A coal-fired compound condensing steam engine powered two reciprocating pumps for a rating of 7,000 gpm. The steel tower familiar to the *Willett* and *Duane* was omitted since the boat had to pass beneath bridges. Named after a former mayor, the *Lawrence* slid the ways at Newburgh, New York, on November 12, 1907, and when placed in service as Engine 87 on December 7, became the first fireboat stationed in the Harlem River. The *Willett* and *Duane,* also named after former New York mayors and placed in service on December 7, were organized as Engine 86 and Engine 85, respectively, both on the North River. These assignments caused a shift in fireboat assignments, which spread coverage of the waterfront, thereby reducing response times.

Ten fireboats strong, the New York Fire Department's fleet was designated by Chief Croker in 1909 as a marine division. Engine company numbers no longer would be assigned to the boats, but to a particular berth.

New York's fireboat fleet periodically was updated to keep pace with development along more than 500 miles of waterfront. The *Boody* was retired in December 1914, and the *Seth Low,* having served the fire departments of both Brooklyn and New York, was replaced on January 1, 1915, by the 118-foot *William J. Gaynor,* which had twice "The Low's" 3,500-gpm output. The *Gaynor* began life as hull number 111 at New Jersey Drydock & Transportation Co. of Elizabethport, New Jersey, in 1913, with delivery the following year. Designers continued the practice of relying on steam turbine-driven centrifugal pumps,

New York's fireboat fleet gathers at the Battery in 1908. From left to right are *The New Yorker*, *Thomas Willett*, *James Duane*, *George B. McClellan*, *Abram S. Hewitt*, *Zophar Mills*, *William L. Strong*, *Cornelius W. Lawrence*, *David Boody*, and *Seth Low*. Author's Collection.

with the only change made in 1937 when the boat's engines were converted from coal to oil-fired. The *Gaynor* went in service on January 1, 1915, as Engine 223 at Brooklyn's Upper Bay.

As WWI entered its third year, New York's waterfront in 1917 was deemed the city's greatest fire hazard due to large supplies of Europe-bound munitions and wooden piers. Fire Commissioner Robert Adamson increased fire protection by organizing an auxiliary fleet of fireboats comprising more than 140 tugboats of New York and New Jersey railroads, Standard Oil, and Bush and Jay Terminal companies. The waterfront was divided into ten sectors and in the event of fire, dispatchers were responsible for locating and sending at least two tugs within that sector. If necessary, tugs dropped their tows at the nearest pier.

The *William J. Gaynor* makes its inaugural cruise on July 3, 1914, before beginning service. Large awnings were common during summer months. In keeping with Chief Croker's preference for ample hose lines for use ashore, three hose reels appear on the trunk deck. The cowl ventilator is polished and a heat shield is forward of the monitor atop the pilothouse. Author's Collection.

The crew of the fireboat *Cornelius W. Lawrence* pauses for a photo. The officer is at lower right. The vessel's relatively short funnel suggests use of a boiler with mechanical draft. Author's Collection.

A rare color image of the *Gaynor* provides a glimpse of its color scheme. Primary colors were white and black, with buff funnel and pilothouse trim. Hose reel covers and decks were gray. Waterways and monitors were red. Builder's plaques traditionally were displayed on the face of the deck house. Courtesy of John Landers-Beth Klein Collection.

The *Zophar Mills,* a product of the late nineteenth century, had served its time on the front lines and was taken out of service in January 1917 at the age of thirty-five years. It remained on the inactive roll until December 1, 1922, when its replacement, the *John Purroy Mitchel,* went in service as Engine 57 at Battery Park. Construction of the 132-foot steel-hull *Mitchel* had gotten underway at Standard Shipbuilding at Shooters Island, New York in 1921. It was New York's last steam-powered fireboat and the only one built as oil-fired. The *Mitchel's* arrival marked a major shift in fireboat assignments, edging out *The New Yorker,* which went to the East River to occupy *Zophar Mills'* berth. The *Mills* was then reactivated as Engine 51 at Staten Island.

A thriving waterfront with ever-increasing marine traffic in New York's waterways kept the marine division busy during the 1920s. With Prohibition imposed during the decade, large numbers of captured "rum runner" vessels served as targets for fireboats to demonstrate and practice with their powerful monitors. Numerous fires during the period ravaged waterfront and ships, testing time and time again the mettle of fireboat crews. At the end of the decade, another name would be added to the list of six men who had died serving aboard New York fireboats. On February 11, 1930, New York firefighting forces worked hard to contain fire aboard the freighter SS *Muenchen* at Pier 42 at the North River. Loaded with highly flammable liquids, the ship exploded, destroying the pilothouse of the nearby *Thomas Willett.* Her pilot, John J. Harvey, was killed and several firefighters were blown overboard. Harvey would not be the last to lose his life in fireboat service, nor would his name fade into obscurity.

In its twilight, nearing the forty-year mark, *The New Yorker* was slated for retirement, but not before its replacement was planned to keep the fleet up to strength. A contract was let in 1928 to naval architect Henry J. Gielow, who became famous for his designs of super-yachts. The contract for building the boat went to Brooklyn's Todd Shipbuilding & Drydock Corp., which began construction in March 1931. Given

The fireboat *John Purroy Mitchel* lies in a quiet channel in 1962. Built in 1921, the 10,000-gpm vessel was named after a New York mayor and replaced *The New Yorker* as flagship of the fleet. The boom attached to the tower was a late addition. It also wears the later color scheme of white and red. Author's Collection.

Tower monitors of the fireboats *Mitchel* (left) and *Gaynor* are used to good effect on fire aboard the British steamship *Egremont Castle* at Brooklyn's Pier 37 on June 24, 1924. Author's Collection.

the age of the fleet's ten steam-powered fireboats, not to mention premature aging of their powerplants from maintaining constant boiler pressure, the vessel was to be powered by modern engines. Despite diesel's growing popularity, it had yet to pass muster with New York Fire Department officials, who chose gasoline engines. Named in honor of John J. Harvey, work began in March 1931 at Brooklyn's Todd Shipbuilding and Dry-Dock Corp. It measured 130 feet long, with a 28-foot beam and 9-foot draft. After being christened by Fire Commissioner John J. Dorman's daughter, Anne, the *John J. Harvey* slid the ways on October 6, 1931. During acceptance trials on December 17, the *Harvey* easily exceeded its design speed of 20 mph

Hose wagons used in conjunction with New York's high-pressure system and fireboats marked the beginning of the end of steamers. The department's first motorized apparatus was this Knox monitor–equipped hose wagon placed in service with Engine Company No. 72 on February 6, 1909. Author's Collection.

Six fireboats of New York's fleet in 1908. Obvious is the variance in pilothouse design, comparing *The New Yorker*'s broad, flat-faced house to higher boot-heel designs of the *Willett* and *Duane* alongside. Author's Collection.

This 1930 architect's drawing, based on designs by the firm Henry J. Gielow, shows the 130-foot fireboat proposed for New York, which would become the *John J. Harvey*. The majority of the trunk deck is occupied by monitor platforms and hose reels. Author's Collection.

Beaming like a proud papa, complete with cigar in hand, Fire Commissioner John J. Dorman shows off New York's newest floating flame fighter, the *John J. Harvey* on December 17, 1931. Officials were aboard for the exhibition, which was part of the acceptance test on the Hudson River. Author's Collection.

and pumped in excess of 18,000 gpm, more than 2,000 gpm beyond its design output. Itwas placed in service later that day as Engine 57 at Battery Park, replacing the *Mitchel* as flagship of the fleet; *The New Yorker* was then retired, having served faithfully for forty years.

Below the *Harvey's* deck, five Sterling Viking II eight-cylinder, 565 hp gasoline engines powered two Westinghouse 1,065 hp propulsion motors to drive twin screws and four centrifugal pumps. The *Harvey* was a Bell Boat with two telegraphs in its pilothouse. The most powerful fireboat of its time, the *Harvey* sported eight monitors, some of which could discharge 3,000 gpm.

When the *Zophar Mills* was retired in 1934 after having served for fifty-one years, the fleet was reduced to nine fireboats and Engine 51 at Staten Island was taken out of service. Lined with piers, warehouses, shipyards, and bulk terminals, New York's more than 250 miles of busy waterfront during the 1930s provided the fireboat fleet plenty of work. Fires on wooden piers and in warehouses, ship collisions and fires, and every imaginable maritime disaster along both New York and New Jersey waterfronts were common occurrences. With the war in Europe growing, the harbors were expected to become even busier, which sparked talk about building a modern and even more powerful fireboat.

Discussions led to plans for New York's next fireboat, which not only followed the high standard set by the innovative *John J. Harvey*, it eclipsed the *Harvey* in power and size. It was a true battleship-class vessel considered king among fireboats in America, and hailed as the most popular. Born of necessity to protect the harbor while a major wartime port, she broke tradition by honoring New York's firefighters, and not a mayor. The *Fire Fighter* was the brainchild of noted naval architect William Francis Gibbs, who proposed to Mayor Fiorello LaGuardia his revolutionary design that mated unrivalled gpm output with the fleet's first diesel-electric propulsion system. Convinced that the unique, powerful vessel would be a shot in the arm for New York City in the wake of the Great Depression, Mayor LaGuardia unhesitatingly approved Gibbs' proposal.

During late 1937 the keel of hull number 856 was laid at United Shipyards at Mariners Harbor, Staten Island. Housed within her 134-foot long, 32-foot wide riveted steel hull were two General Motors 1,500 hp, sixteen-cylinder Winton diesel engines turning a pair of 6-foot-diameter, three-blade bronze screws with 4-foot pitch; Gibbs included propulsion control and electro-hydraulic steering of two rudders from the pilothouse. Four De Laval 5,000-gpm two-stage

With flying pennants used on special occasions, the *John J. Harvey* sails from the Battery on January 16, 1938, as fire commissioner John J. McElligott hosts officials of the International Fire Chief's Association. Hose reels and monitors left little room on *Harvey's* deck. The high antenna at the pilothouse was for radios, introduced to the fleet in October 1937. Author's Collection.

Displaying a subtle sheer, the *Harvey* appeared as a steam-powered vessel but was gasoline-powered, producing more propulsion power than steam during pump operations. Bronze name boards, or trailboards, were mounted on the bow, pilothouse, and stern. Author's Collection.

centrifugal pumps drove four Westinghouse Marine 600 hp motors that delivered the *Fire Fighter's* 20,000 gpm punch through nine monitors. Most powerful was the bow monitor, which could discharge 27 tons of water per minute, allowing it to also function as a formidable tool of destruction. Another monitor, rated at 3,000 gpm, topped a unique hydraulically operated tower abaft the deck house. As a floating pumping station, *Fire Fighter* boasted two manifolds with twenty-four outlets, along with three large hose reels. Included in its extensive inventory were four 500-gpm foam generators, which were later replaced

Replete with contrasting paint schemes, polished metal fixtures, and detail-painted hose reels, New York fireboats were worthy of admiration. At the dock in 1938, the crew of the *Harvey*, in dress uniform, beam proudly from the pilothouse. Both vessels have hose-fed nozzles mounted to their bulwarks. Author's Collection.

by a modern foam system. Fully loaded, it displaced nearly 600 tons, yet could attain a top speed of 16 mph.

The honor of christening the *Fire Fighter* went to Eleanor Grace Flanagan, the daughter of a New York fireboat lieutenant, for attaining the highest scholastic average of all fire department children in city high schools. On August 28, 1938, young Ms. Flanagan smashed a champagne bottle against the vessel's bow, and at the moment the mighty *Fire Fighter* slid the ways into Newark Harbor, United Shipyards went out of business; the yard's new owner, Bethlehem Steel, would outfit the fireboat. Before beginning her watch, the *Fire Fighter* was used to train crews in its substantial capabilities while berthed at department shops on the Hudson River. On November 16, it reported for duty at Engine 57 quarters at the Battery. That bumped the *Harvey* from Engine 57 flagship status to Engine 86 at pier 53 North River, which saw the *Willett* transferred from Engine 86 to Engine 78 on the East River, which, in turn, moved the *McClellan* from Engine 78 to Engine 66 at Grand Street, East River. The *Strong* then gave up its spot at Engine 66 to become reactivated Engine 51 at Staten Island.

The *Fire Fighter's* first run came just three days into its active service for an oil barge adrift on the Hudson, and it fought its first fire on January 23, 1939, at Brooklyn's waterfront, supplying thirty hand lines for a blaze aboard the British freighter *Silver Ash,* laden with rubber and cyanide.

After basking in the limelight at the 1939 World's Fair, the *Fire Fighter,* along with its sister fireboats, returned to the serious business of protecting a port

The *Thomas Willett* and crew battle a fire at Pier 54, North River on May 6, 1932. Less than two years earlier, the *Willett's* pilothouse was destroyed when a burning ship exploded, killing fireboat pilot John. J. Harvey. During the 1940s, a platform with two monitors was added amidships. Author's Collection.

The *Willett* puts out nine streams at the celebration of the opening of the Triborough Bridge to vehicle traffic on July 11, 1936. The bridge was one of the largest public works projects of the Great Depression, more costly than the Hoover Dam. Recent modifications to the *Willett* seen here included a radio mast affixed to the pilothouse, and a monitor platform amidships. Author's Collection.

that began teeming with cargo ships supporting the war in Europe, and sleepy shipyards that came alive with warship production. With military ships of every description, and transports laden with fuel, ammunition, and cargo for the war effort crowding the harbor, the potential for disaster was overwhelming. Fires in warehouses and pier houses overloaded with war materiels often burned, with disastrous results. Noteworthy among the blazes fought by the *Fire Fighter* and other New York fireboats during WWII was the fire that consumed the oceanliner *Normandie* during conversion to a troopship at Manhattan's Pier 88 on February 9, 1942. When the mooring lines of the burning liner snapped and it began to capsize from the huge amount of water pumped aboard, the *Fire Fighter* was almost crushed as the liner rolled onto its portside.

The following year, the *Fire Fighter* and the *John J. Harvey* joined forces on April 24 to battle fire aboard the *El Estero* as aviation fuel was being loaded at the army's Caven Point Terminal near Jersey City. With 1,400 tons of explosives already aboard and the ship berthed among storage tanks filled with millions of gallons of petroleum products, the stage was set for disaster. Since the 27,000 gallons of water per minute from the two fireboats had little effect, it appeared certain the *El Estero* would explode and cause devastation throughout the New York and New Jersey port region. Residents were warned to prepare

Firefighters from the Gaynor board the heavily listing SS *City of Montgomery* on March 6, 1935. The ship caught fire off the Atlantic coast and was able to make it to shore. The fireboat *Willett* is alongside. Author's Collection.

At the base of the *Fire Fighter's* monitor tower, seen here lowered, was a 50 hp air compressor for operating pneumatic tools such as air hammers for breaching piers. The tower proved mechanically troublesome and was removed in 1962. It was raised hydraulically 55 feet above the waterline to bring to bear its 3,000-gpm monitor. Courtesy of John Landers-Beth Klein Collection.

Since nearly all of the fifty-one nozzles carried aboard the fireboat were made of brass, the room in which highly polished nozzles and firefighting appliances were kept was called the Gold Room. Courtesy of Morris Gordon.

The FDNY Marine Division assumed full use of Manhattan's Pier A, Battery Park in 1960. The Victorian structure, which incorporates a four-story clock tower donated by US Steel founder Daniel G. Reid, was placed on the National Register of Historic Places in 1975. Both the structure and the fireboat *Fire Fighter* have become iconic figures on the city's waterfront. Author's Collection.

for the worst. Flanked by the *Fire Fighter* and the *Harvey,* the burning freighter was towed out to sea, where she finally sank under the weight of water from fireboat streams. The two fireboat crews, along with Jersey City firefighters, were cited for heroism in averting certain disaster. Such heroism was commonplace. Logbooks of FDNY fireboats during WWII and during the postwar era testify to the mettle of their crews as they fought countless major fires aboard ships and along New York and New Jersey waterfronts.

When the *Strong* was removed from service in May 1948, Engine 51 was without a fireboat until September 1949, when the *George B. McClellan* took up station at Staten Island. Disaster struck the *McClellan* on September 17, 1953, when its boiler exploded, resulting in the death of marine engineer John D. McKean. The 1950s saw a marked increase in major pier fires, culminating with the catastrophic fire and explosion on December 3, 1956, at Luckenbach Pier. Authorities cried "enough!" and pier operators faced strict regulations outlining fire prevention measures. During these heydays of the battleship-class fireboats, a maximum number of crewmen were assigned; the crew of the *Fire Fighter*, for example, comprised pilot, two engineers, wiper, fire officer, and four firefighters. Additional firefighters went aboard as the situation dictated. The 1950s heralded sweeping changes in the FDNY Marine Division, which had been reduced to eight fireboats with retirement of the older vessels. At the beginning of the decade, no immediate

Largest among the *Fire Fighter*'s monitors is the bow monitor, which can deliver 27 tons of water a minute through a 5-inch-diameter tip. The deck above the oversize pilothouse was extended aft to the funnel to form a multiple monitor platform. Black-painted hose reel covers originally were polished brass. Courtesy of John A. Calderone.

Eight firefighting and assistance vessels crowd the port side of the burning liner *John Ericsson* at New York's Pier 90 on March 7, 1947. The *Queen Elizabeth*, whose destiny with fire would come, lies in the background. Author's Collection.

replacements were in store; however, 1952 plans for a fireboat for New York City were found among records salvaged from the John H. Mathis Company of Camden, New Jersey. Mathis was an old name in the business, having years of experience building ships for peace and war. Although the plans had been tabled, one thing was certain: the next fireboat to join the fleet would immortalize John D. McKean. The mighty *McKean* did, in fact, begin life in 1954 at the Mathis yard as hull number 206. Nearly as large and as powerful as the *Fire Fighter*, the *McKean* was of battleship-class in every respect. Designed by New York's John H. Wells, it displaced 330 tons and measured 134 feet long, with a 30-foot beam.

Advances in mechanical engineering allowed designers to deviate from the tried and true practice of using the same powerplants for propulsion and pumping. Housed within the *McKean's* hull were two 1,000 hp supercharged, reversible diesel engines driving two screws recessed into tunnels, and two 1,000 hp pumping engines powering four 4,500-gpm Worthington pumps. All four engines were built by the Enterprise Engine and Machinery Company. The pumps could deliver 19,000 gpm to six 3,000-gpm monitors, one of which was mounted atop a 39-foot tower, placing it 48 feet above the waterline. Built into the deckhouse were sixteen discharge outlets. Three hose reels, one forward and two aft, held 4,500 feet of large diameter hose.

The diesel-electric *John D. McKean* was christened on March 19, 1954, by Susan Wagner, wife of Mayor Robert F. Wagner, who later became senator and had an FDNY fireboat named for him. The *McKean* was commissioned on May 1, 1955, beginning its watch as Engine 57 at Battery Park, sending the *Fire Fighter* southward to become Engine 223 at 37th Street, Brooklyn, Upper Bay, which is the heart of the Port of New York and New Jersey.

The *George B. McClellan* in 1948 wearing the fire department's original color scheme for fireboats. Its fire service career ended with the explosion that claimed the life of marine engineer John D. McKean on September 17, 1953. It was disposed of the following year. Courtesy of John Landers-Beth Klein Collection.

Concern about the hazards of the quarter-century-old *Harvey's* gasoline engines sent it to overhaul, which included refitting with safer, more cost-effective diesel engines. The installation of five Fairbanks-Morse diesels was engineered by Henry Gielow, the boat's original designer, with the job contract awarded to Camden's RTC Shipbuilding. Updates included a 3-foot extension to the deckhouse, which encased the lower sections of two new funnels. Overhaul was completed in 1957 and the *Harvey* was back at its Bloomfield Street berth on the North River.

One of the changes that occurred during the period was the designations in March 1959 from engine companies to marine companies. Marine companies numbered 1 through 9, with Marine Co. 4 reserved for future use. That year also marked the beginning of a program to repaint the black hulls of fireboats red, with bulwarks and superstructures remaining white. Funnels were changed from buff with black top sections to red. Trim work, including taffrails, pilothouse visors, funnel caps, and hose reels, were black.

With the retirement, and impending retirement, of seven fireboats during the 1950s and 1960s, the FDNY laid plans for modernizing the fleet. The *McClellan* was retired after the fatal explosion in its engine room in 1953, and the *Lawrence* was retired in 1955. The *Hewitt* was relegated to Staten Island's ship graveyard in 1958, and the *Willett* and *Duane* were retired in 1959, the latter being sold to Ships & Power of Miami in September 1961; that year also marked disposal of the *Gaynor*. The *Mitchel* was New York's last remaining steam-powered fireboat, ending forty-five years of service in February 1966. Compared to its proud and distinguished fire service career, the *Mitchel's* twilight years were, at best, ignoble. After being sold at auction in 1966, a group representing the buyers showed up at Marine Division Headquarters with road maps of eastern coastal states from New York to Florida. Although FDNY crewmen suggested they obtain nautical charts to take the boat southward, the group assured officials that they would be okay since they would hug the coast and remain in sight of land. The inexperienced group got as far as Delaware, where they ran the boat aground. It got worse; the new owner decided to swim ashore for help and drowned. The former *Mitchel* eventually made it to the Gulf, where it was briefly put to use as a shrimp boat, but also was used surreptitiously to dump toxic chemicals into the ocean off the Texas coast. After a drum of the toxic mix washed ashore, the boat was seized by Texas authorities and the owners arrested. It was purchased at government auction by buyers bent on turning the vessel into a floating restaurant, but the plan fell

Designated Marine Company No. 1, the *John D. McKean* long reigned as flagship of the fleet. Built in 1954, the 19,000-gpm vessel operated with a crew of seven. Courtesy of John A. Calderone.

Marine Company 1, the *McKean*, uses three monitors to wet the remains of a massive waterfront blaze. Courtesy of Ron Frehm.

The *John D. McKean* in its element at a seven-alarm fire at Brooklyn's Pier 39 on April 6, 1973. The firefighter at the bow is dwarfed by the massive 7,500-gpm monitor. The monitor tower had been rebuilt from its original configuration. Courtesy of John A. Calderone.

through and the former fireboat ended up eroding in a Texas marsh with the *John Purroy Mitchel* nameplate still on its pilothouse.

Modernization of the fleet was accomplished with the purchase of four new fireboats built in the Mathis yard, birthplace of the *McKean,* setting an unprecedented record of turning out five battleship-class fireboats. They were named the *H. Sylvia A. H. G. Wilks, Harry M. Archer M.D., Senator Robert F. Wagner,* and *Governor Alfred E. Smith.* The four were identical, with the exception of manually operated monitors on the *Smith,* along with a differently shaped funnel. All were 105 feet, 6 inches long, with a 27-foot beam and 9-foot draft. Power was derived from a pair of 500 hp diesel engines turning two variable-pitch screws, and two 500 hp diesel pumping engines delivering 8,000 gpm. Monitors were hydraulically operated and remotely controlled. The four sister ships each were fit with a 500-gallon foam tank.

The *Wilks* got its convoluted, non-traditional name from Hetty Sylvia Ann Howland Green Wilks, the daughter of Henrietta Howland Robinson Green, also known as the "Witch of Wall Street." Sylvia endowed much of her inherited wealth to funds, one of which benefited New York City firefighters and their families. Another connection with the FDNY was Sylvia's brother, Edward, who for many years was a buff at lower Manhattan's Hook & Ladder Co. 3.

First among the four new fireboats to slide the ways at Mathis' yard was hull number 211, the *Wilks,* on January 16, 1958, under sponsorship of Fire Commissioner Edward Cavanaugh's wife. It began service on May 13 as Engine 77 at Brooklyn's Fulton Street on the East River; this would become Marine Co. 7 nine months later.

Next was hull number 212, the *Archer,* named after Dr. Harry M. Archer, who, during his early years, had set up treatment facilities at disasters. He served as honorary surgeon of the FDNY and was named second deputy fire commissioner before his death in 1954. After christening by Archer's daughter, the

The fireboat *John H. Glenn Jr.* and the tender *Smoke II* support dive operations in 1963. The *Glenn* had just arrived and *Smoke II* was soon laid up for rebuild, having its long cabin removed and pumping capability increased. It was FDNY's fourth fireboat tender. Courtesy of John Landers-Beth Klein Collection.

The 9,000-gpm *H. Sylvia A.H.G. Wilks* more than a decade after it began service as Engine Company No. 77 at Fulton Street and the East River in 1958. Courtesy of George F. Mand Library, FDNY.

The *Wilks* was the first FDNY fireboat equipped with remotely controlled, hydraulically operated monitors. A helm on the afterdeck allowed steering when rearward visibility was necessary. Staten Island ferries are moored in the background. Courtesy of John Landers-Beth Klein Collection.

The pilothouse (also called wheelhouse) of the *Wilks* when it was new in 1958. Flanking the helm, or wheel, at far ends are port and starboard engine throttle pedestals. Purposely positioned directly forward of the wheel at waist level, in line with the helmsman's vision, is the nautical compass binnacle. Immediately to its right is an engine room telegraph pedestal. At far right is the pump engine telegraph. Courtesy of John Landers-Beth Klein Collection.

fireboat began service as Engine Co. 66 at Grand Street, Corlears Hook, East River, which became Marine Co. 6.

The *Wagner,* hull number 215, was named for Senator Robert F. Wagner of New York, who was instrumental not only in labor reform, but in safety legislation in the aftermath of New York's disastrous Triangle Shirtwaist fire on March 25, 1911. The *Wagner* began service on December 8, 1959, as Marine Co. 5 at East 90th Street on the East River.

The daughter of Dr. Harry M. Archer christens one of four new large fireboats after her father. For the arrival of the *Harry M. Archer M.D.,* Marine Company 6 was organized in 1959 at Grand Street, East River, Manhattan. Courtesy of John Landers-Beth Klein Collection.

Hull number 216 was named after four-term New York Governor Alfred E. Smith, who championed labor and fire safety reform in the wake of the Triangle Shirtwaist Factory fire. The *Smith* was placed in service as Marine Co. 8 at Brooklyn's Upper Bay on the anniversary of Governor Smith's death, October 4. Shortly after the *Governor Alfred E. Smith* left the John H. Mathis yard in 1961, the company closed its doors.

The period also marked the arrival of a new tender to replace the *Smoke,* which retired in 1955. Named *Smoke II,* the tender's design was based on rugged boats that served oil-drilling rigs in the Gulf of Mexico. Equitable Equipment Co. of New Orleans built *Smoke II* with a steel hull housing two 300 hp General Motors diesel engines, giving the boat a top speed of 24 mph. A roomy enclosed cabin enabled transport of personnel or accommodations for four litter patients. Weighing 34 gross tons, it measured 52 feet long, with a 14-foot beam, and its 4-foot draft allowed its crew of two to navigate shallow waters. An Aurora 250-gpm centrifugal pump gave the *Smoke II* limited firefighting capability, along with a built-in mechanical foam system with 80-gallon foam tank. *Smoke II* was rebuilt during 1963 and in the late 1970s its large enclosed cabin was removed, and two centrifugal pumps were installed to deliver 2,000 gpm through monitors and hose outlets.

The pilothouse of the *Governor Alfred E. Smith*, which began service as Marine Company 8 in 1961. Control layouts in the wheelhouses of the big boats were fairly consistent. Above the steam radiators on the forward wall are electric meters. Courtesy of Britton Crosby.

The *Mitchel* was the FDNY's last remaining steam-powered fireboat, which went into retirement in 1966. Courtesy of John Landers-Beth Klein Collection.

The *Governor Alfred E. Smith* performed frontline service until being placed in reserve in 1994. Brass bells and name plates mounted forward on the deckhouse was a time-honored tradition. Courtesy of Britton Crosby.

Details of one of the *Wilks'* remotely controlled monitors at work. Courtesy of John Landers-Beth Klein Collection.

The fireboat *Harry M. Archer M.D.* and crew fight fire aboard the containership *Sea Witch* after it rammed the oil tanker *Esso Brussels* in New York Harbor on June 2, 1973. The collision touched off an explosion that set both ships ablaze and killed at least ten crewmen. Courtesy of George F. Mand Library, FDNY.

Small boats found their place among the jumbo fireboats of the FDNY during the 1960s. The largest of the small boats was the 29-foot *James E. Hackett*, named after a deputy fire commissioner. The boat was assigned to Marine Company 9, the *Fire Fighter*, as a satellite. Four years later, in 1974, it was given to the business community of City Island, a small island of the Bronx at the far end of Long Island Sound. Courtesy of John Landers-Beth Klein Collection.

The FDNY joined the trend in small waterjet-powered boats with this 20-footer named *Blaze*. A 600-gpm pump supplied a bow monitor. Courtesy of John Landers-Beth Klein Collection.

During the 1960s, the face of marine operations in the New York and New Jersey harbor regions changed once again. The number of traditionally built wooden piers was diminishing, and the transport of hazardous materials by ship not only was minimized, but came under strict regulations. These changes, along with containerization of bulk cargoes, eliminated the need for a fleet of large fireboats. Battleship-class fireboats still had their place, but there arose a need for smaller, faster vessels to negotiate growing marinas, deal with pleasure craft, and perform rescue. Advances in technology allowed smaller boats to do more with less, which translated to less manpower, which, in turn, eased budgets. Accordingly, the FDNY joined the trend in small waterjet-propelled conversions of commercial hulls, but for seasonal use from May to October. A 20-footer named *Blaze* came from the Mathis yard shortly before the firm ceased operations in 1961. Outfitted with a 600-gpm pump and bow monitor, the *Blaze* was designated a special unit and is believed to have been assigned initially to newly organized Engine 329, Fort Tilden, Queens, for ocean beach patrol. The boat was then experimentally assigned to Marine Co. 8, the *Smith.* Although a small boat, the *Blaze* proved cumbersome for frequent transfer by davit to the water; it was assigned to Queens Engine 331 until removed from service in December 1970.

Painted white, the *Blaze* was experimentally assigned to Marine Company 8, the *Smith*, but proved too cumbersome for deployment from the fireboat. The boat was assigned to a land engine company and eventually removed from service. Its midship-mounted waterjet supplied gated valves at the rear deck. Courtesy of John Landers-Beth Klein Collection.

In 1963, the FDNY acquired an 18-foot, waterjet-propelled boat with a 680-gpm pump and monitor to operate as an element of Engine Co. 329 at Rockaway Beach. Named the *Flame,* the boat remained a seasonal fixture along the coast until 1974.

To cover the upper reaches of the East River where it widens into Long Island Sound, the FDNY in 1962 purchased a fireboat that not only was smaller than the big boats in the fleet, but was faster and of shallow draft to maneuver among small boats. Naming the boat was the most significant deviation from the custom of selecting notable namesakes connected with New York. It was named *John J. Glenn Jr.* in honor of the first American astronaut to orbit the earth, a blockbuster event that occurred while the fireboat was in design stage, making it the first FDNY fireboat named for a living person.

The *Glenn* was designed by New York's H. Newton Whittelsey and built by Diesel Shipbuilding Company, Jacksonville, Florida. Displacing 81 tons, it measured 70 feet long, with a 21-foot beam and 5-foot draft. Three 450 hp General Motors diesel engines turned three screws and powered two centrifugal pumps to deliver 5,000 gpm through four monitors. The *Glenn* was placed in service on September 13, 1962, as newly organized Marine Co. 4 at Fort Trotten in the borough of Queens. Its area of operations included northern Queens, southern and eastern shores of the Bronx, and numerous islands.

The span of ten years between the mid-1960s and mid-1970s marked major changes along the waterfront, and subsequent changes in the fireboat fleet. As fewer major fires occurred along the waterfront, changes in port operations and expanding marinas required increased fire and rescue services for pleasure boaters. The concept of smaller boats never seemed more sound as when the budget axe took its mightiest swing at the FDNY, cutting the fireboat fleet in less than half with the elimination of five marine companies. Left was Marine Co. 1 the *McKean,* Marine Co. 2, the *Harvey,* Marine Co. 6, the *Archer,* and Marine Co. 9, the *Fire Fighter.* The *Fire Fighter* in 1967 relocated to the Homeport Pier district at Tompkinsville on the northern tip of Staten Island. The *Wagner* and the *Smith* were placed on reserve status, the *Mitchel* was retired in 1966, the *Wilks* was transferred in 1972 to the Division of Marine and Aviation as a shop tug, and in 1977 the *Glenn* was sold to the Washington, DC, fire department. The *Harvey* was taken out of service during the late 1970s, overhauled, and returned to its Pier 53 berth on the North River in 1981.

The mid-size 5,000-gpm *John H. Glenn Jr.* began service in 1962 as newly organized Marine Company 4 to cover upper reaches of the East River. The *Glenn* is seen here in its original configuration. Courtesy of John Landers-Beth Klein Collection.

With the exception of small service boats, no fireboats would be purchased during the thirty-year span from 1962 to 1992. Largest among the small boats was the 29-foot *James F. Hackett,* acquired in 1970 as a satellite for Marine Co. 9, the *Fire Fighter.*

Although measuring only 18 feet long, the *Flame* was equipped with a 680-gpm pump and monitor to cover the Atlantic beachfront. The boat was assigned to Engine 329 at Rockaway Peninsula, Queens. Courtesy of John Landers-Beth Klein Collection.

Only 8 feet wide and drafting 2 feet, 6 inches, the fiberglass boat delivered a respectable 2,500 gpm at 120 psi, making it ideal for work in confined areas. The *Hackett* saw only seasonal use and was removed from service in 1974. The boat was sold to the City Island business community for a token $1; however, repairs proved too costly and the boat remained idle.

Few changes were made to the *Fire Fighter*, which stood vigil at the entrance to New York Harbor, where numerous ships lightered their cargo onto smaller ships. That's what the oil tanker SS *Esso Brussels* was doing on the night of June 2, 1973, when the containership SS *Sea Witch* lost steering and slammed into the *Brussels*. Thousands of barrels of burning oil

The 8,000-gpm *Harry M. Archer M.D.* in 1967. Although launched with a black-painted hull in 1958, its hull shortly thereafter was re-painted red in keeping with new color scheme directives for the fleet. The *Archer* was the second launched of four sister fireboats for the FDNY, serving until 1994. Courtesy of George F. Mand Library, FDNY.

The Superpumper System's original trio of Satellites was Mack Model C85s carrying 2,000 feet of 4½-inch hose and mounting a 4,000-gpm Stang Intelligiant monitor with 6-inch dimeter tip. From 1965 to 1982, the Superpumper System worked at more than 900 fires. Courtesy of John A. Calderone.

Many years before noted naval architect William Francis Gibbs designed the fireboat *Fire Fighter*, he envisioned a "land fireboat." Technology caught up with his idea in the mid-1960s with FDNY's Superpumper System, comprising pumper, this tender and three satellite trucks. The massive Superpumper could pump nearly 9,000 gpm. The stream of the tender's 10,000-gpm monitor reached 600 feet, and was capable of reducing concrete to rubble. Courtesy of Chuck Madderom.

Replacing the Superpumper System by 1983 was the Maxi-Water System, which in 1999 became the Satellite Water System consisting of six satellites and six 2,000-gpm pumpers. Satellites were basically hose and foam wagons mounting a fireboat-type monitor, which could be supplied water from multiple sources, including fireboats. Satellite 1 was a 2002 Mack MR/Saulsbury truck. Courtesy of John A. Calderone.

spread flames 1,000 feet across the water; hundreds of cargo containers blazed on the *Sea Witch's* main deck. The *Fire Fighter's* crew of ten wasted no time rushing to the hellish inferno, which was searing the nearby Verrazano Bridge. Locked in death throes, the two blazing ships drifted on the outbound tide, the *Fire Fighter* following with all monitors thundering away. When the burning pair ran aground at Gravesend Bay, firefighters were astounded to see a flashlight blinking in desperation from the *Sea Witch's* fantail. Using the bow monitor to clear a path through oil burning on the water, *Fire Fighter's* crew closed the distance to see scores of crewmen huddled on the fantail, trapped by heat and smoke. Despite searing flames that turned the *Fire Fighter's* foredeck hot, firefighters remained at their monitors while the pilot nosed against the ship, allowing firefighters to raise ladders to rescue thirty-one souls. The following year, *Fire Fighter* and her crew received Department of Commerce and Merchant Marine honors never before awarded a fireboat and crew.

Such examples of skill and courage ensured that the *Fire Fighter's* usefulness was never questioned, despite its age and size. To prolong that usefulness, during the late 1970s it went to Direcktor Shipyards at Coddington Cove, Rhode Island, for complete overhaul. Back on the front lines, it continued to show its worth, which included pumping for thirty continuous hours at a fire and explosion on a hazardous materials site on Staten Island.

The New York Port Authority has been a partner organization to the FDNY since its creation in 1921. Its aviation assets, mainly helicopters, along with those of the NYPD, prove beneficial to FDNY marine operations with the common purpose of safeguarding airports and ports of New York and New Jersey. One of Port Authority's first helicopters was this 1949 Bell Model 47D1 seen here at LaGuardia Airport in 1952. Courtesy of Frederick G. Freeman.

The fireboat tender *Smoke II* in its final configuration following removal of the large cabin, and wearing the FDNY's modern paint scheme. Courtesy of John A. Calderone.

When Brooklyn Box 55-4137 was struck on May 10, 1962, at Sinclair Oil Terminal, Mill Basin, firefighting forces faced 100,000 gallons of burning gasoline. The inferno instantly rose to eight alarms, a Brooklyn 5th alarm and a Queens 3rd alarm. The fireboat *Governor Alfred E. Smith* wages war with the blaze. Courtesy of George F. Mand Library, FDNY.

Having celebrated its fiftieth year of service, the venerable *Fire Fighter* in 1989 was declared a National Landmark and placed on the National Register of Historic Places. As fit as when new, and in nearly original state, it continued its reign as the most powerful fireboat in the fleet. New York Fire Department officials seemed to have no intention of replacing it.

Encouraged, perhaps, by the success of Tacoma, Washington's Surface Effect Ship (SES) fireboats, FDNY officials during the late 1980s set their sights on American-built surface effect ships to enhance coverage of the waterways. In addition to firefighting, it was intended that the SES vessels be used for search and rescue, security and patrol of the harbor, and pollution monitoring. Plans were made from which a pair of SES fireboats materialized, but the New York Fire Department would not share the success enjoyed by the Tacoma Fire Department.

After awarding the contract for the vessels in mid-1990 to Textron Marine Systems of New Orleans, FDNY marine officials worked with the firm to design and build the boats. Textron Marine Systems was an outgrowth of Bell Aerospace Textron, whose New Orleans division had teamed with Halter Marine Services to develop an SES. Prior to venturing into production of the first American-built surface effect ships, Halter had specialized in military landing craft and Coast Guard motor lifeboats. In the end, the general consensus was that the firm should not have veered from its line of military craft.

Basically, the SES system uses a cushion of air trapped between catamaran-type sidehulls and flexible bow and stern seals to lift the center hull section. The air cushion is created by diesel-blower engines. Overcoming resistance with the water allows great speed and the ability to operate in shallows. Twin screws of each 70-foot fireboat were powered by two 460 hp General Motors marine diesels for a speed of 35 mph, while another, rated at 400 hp, powered the lift fan, and a fourth powered three centrifugal pumps to deliver 7,075 gpm at 100 psi. The beam of their glass-reinforced plastic catamaran hulls spanned 20 feet. While idle they drafted 5 feet, and a mere 3½ inches when underway.

Being smaller, faster, and nearly void of draft restrictions, the two SES boats were to be teamed with battleship-class fireboats at outlying berths, allowing them quick access to farthest reaches of city waterways. Incorporating the pair into the fleet would involve a major shift in fireboat location. Since city officials had long eyed the historic Pier A at the Battery for private development, Marine 1, the *McKean*, would move to new quarters at the Brooklyn Navy Yard on the East River. Marine 2, the *Harvey*, stood fast at Pier 53, North River, as did Marine 9, the *Fire Fighter*, at Staten Island. Marine 6, the *Archer*, would occupy new quarters under construction at the FDNY Training Academy on Randall's Island. The SES boats would berth with the *Archer* at the northern end and with the *Fire Fighter* most southern.

While the SES boats were under construction, along with a smaller, more conventional fireboat being built at a Massachusetts yard, it was announced in spring 1992 that the three would be named for the last three firefighters killed in the line of duty. They were: John P. Devaney, Ladder Co. 131, killed February 3, 1989; Alfred E. Ronaldson, Rescue Co. 3, killed March 5, 1991, and Kevin C. Kane, Ladder Co. 110, killed September 13, 1991.

The 52-foot, 5,000-gpm fireboat *Kevin C. Kane*, after beginning service at new Brooklyn quarters in 1992, claimed the title of "busiest fireboat in the world." It was named in honor of a New York firefighter killed the previous year. Courtesy of John A. Calderone.

No longer recognizable as the FDNY fireboat *Archer* disposed of in 1994, the vessel was transformed by Crescent Towing into the tug *Miriam Walmsley Cooper*. The firm had also acquired the fireboat *Wagner*, which was stripped to its hull and rebuilt as the modern tug *Georgia*. Courtesy of Jim Murray.

The fireboat *Governor Alfred E. Smith* was the last of four sister ships built at the Mathis Shipyards at the beginning of the 1960s. Upon the *Smith's* arrival, the fireboats *Duane* and *Gaynor* were retired. Here, the *Smith* approaches the Brooklyn Bridge in 1987. Author's Collection.

The SES boats *John P. Devaney* and *Alfred E. Ronaldson* arrived for tests in 1992, but a host of technical problems delayed their operational use until late June 1994. After less than four months of service, the boats were deemed a failure and removed from service. Tradition dies hard in the fire service and many felt from the onset that the SES boats were unworthy of bearing the names of firefighters who paid the ultimate price.

Besides the SES boats, other radical changes in the FDNY's fireboat fleet had been planned early in the decade. By 1990, internal opposition and the loss of political and fire department officials that supported the purchase of amphibious fireboats brought a quick end to the venture. Louis DeChime, who had brought the military amphibian-turned-fireboat concept to fruition, was poised to aid in the construction of a handful of the unique vessels for FDNY use.

Events leading up to the proposed change in New York fireboats had begun during the late 1970s when NASA and the US Coast Guard embarked on a long-term development project aimed at mating surplus military LARC amphibians with powerful, high-tech turbine engines and pumps. After new owners of the firm LTI ceased production of amphibious fireboats in 1988, Cunningham Marine Hydraulics stepped in, striking a deal with DeChime and partner Robert Milnes of Fire Fighting Innovations to revive the concept. Robert "Fritz" Cunningham, who had established his firm in Hoboken, New Jersey, in 1976, was no stranger to the LARC, having converted two of the amphibians to total hydraulic drive. The pair was used by the Army Corps of Engineers for survey work in New York City waterways.

During the same period, New York City Fire Chief Homer G. Bishop, who was fully aware of the success of Miami's, Milwaukee's, and Tampa's amphibious LARC fireboats, sought alternative designs for FDNY's next generation of fireboats. Besides surface effect vessels, amphibious fireboats placed high on his list.

The unique but unsuccessful Surface Effect Ship (SES) fireboat *John P. Devaney* passes Manhattan's storied skyline during sea trials in the early 1990s. After less than four months of service in 1994, the *Devaney*, along with its sister SES fireboat, the *Alfred E. Ronaldson*, were deemed failures and removed from service. Courtesy of Ron Frehm.

One of two US Army Corps of Engineers LARCs converted to hydraulic drive by Cunningham Marine Hydraulics of Hoboken, New Jersey, and used for fireboat feasibility tests at New York City's City Island. With officials aboard, the LARC V heads down City Island's boat ramp to begin water evaluation. Courtesy of Bob Milnes.

Especially appealing was the cost-effectiveness of the design, which allowed it to be stored in a fire station, maintained like a land vehicle, and cross-manned by existing station personnel; a crew of three was sufficient for its operation. Nor was it lost on Bishop that the amphibian reduced response times and was highly maneuverable and stable in rough waters.

Opposition, however, loomed when the FDNY marine division and the union expressed misgivings about fireboats with wheels kept in fire stations coming under the control of the operations division rather than the marine division. Nevertheless, Chief Bishop saw merit in the design, particularly for City Island, which had a boat ramp near the fire station. He did, however, request that the pump control panel be modified to a conventional type, and that the cabin be altered to accommodate seating for six firefighters. Bishop planned two amphibians, with a follow-on order for between three and five for other locations. The fire department bid on the design, with the option of tag-on for the second vehicle, but the bid was rejected due to the intensifying internal power struggle.

Cunningham complied with Bishop's redesign, and DeChime and assistant Jerry Winskie began fabricating a full-size mockup, using Coral Gables' LARC they had purchased at auction. After Chief Bishop inspected and approved the mockup design, it went to Cunningham's Hoboken facility for construction. In the meantime, one of Cunningham's hydraulically reconfigured LARCs was tested at City Island for space restrictions on streets and at a fire station. Proven to be a good fit, and easier on the budget, the unique amphibian was sure to silence opposition. Sold on the concept, Chief Bishop assured a re-bid and urged that construction begin immediately to ensure its debut at forthcoming Fire Prevention Week events, which could only help still the now-fierce opposition.

The FDNY version of the amphibious fireboat—the amazing new addition to the fireboat fleet that had the potential to restructure the thinking in fireboat design—all aluminum, with new hydraulics, a 700 hp 12-cylinder engine, custom radiator, and three hydraulic

One phase of the LARC's feasibility tests was checking for fit in a City Island firehouse. Courtesy of Bob Milnes.

Road testing during evaluation of the LARC as an amphibious fireboat demonstrated its tight turning radius on land. On water, the LARC's front wheels were used in conjunction with a rudder to negotiate tight quarters. Courtesy of Bob Milnes.

Windy was a LARC V intended for use by the Coral Gables, Florida, fire department as a rescue vehicle during floods caused by hurricanes. Louis DeChime and his assistant used the amphibian to fabricate the mock-up to meet FDNY specifications. The final structure would be all-aluminum. Courtesy of Bob Milnes.

Scale model showing the superstructure and main monitor configuration of LARC amphibious engines proposed for the FDNY. Courtesy of Louis DeChime.

Elkhart monitors—was never ordered. Confident that orders were forthcoming, and with Buffalo; Jersey City; San Juan, Puerto Rico; Guam, and Kuwait fire departments, along with Malaysia's Civil Aviation Authority, waiting in the wings for the nod from New York City to place orders, DeChime and partners faced a dramatic turn of events. In 1989, New York City Mayor Ed Koch was defeated in his bid for re-election, Chief Bishop took ill and passed away in 1990, and other FDNY supporters of the plan retired. Chief Bishop's successor, who was said to be under political pressure to cut costs, disapproved the amphibious fireboat from the onset, sealing its doom.

Cunningham sold the LARC and components and shut down the project. DeChime summarizes, "I was cured of insanity and dropped the amphibious fireboat dream forever, a thirteen-year obsession."

While discontinuing use of the SES boats cancelled planned berth changes, other noteworthy events impacted the fireboat fleet. In 1991 Marine Co. 2 was disbanded, although the *Harvey* would serve in reserve status until officially retired in 1994. That year also saw disposal of the *Archer,* and the *Smith* assuming sole reserve status. Marine Co. 1, the *McKean*, moved to the *Harvey's* former berth at Pier 53.

Emerging successfully from the SES and amphibious fireboat experience was the only fireboat built for the FDNY during 1992: the *Kevin C. Kane,* which began service as Marine Co. 6 on September 27, 1993. The 6,500-gpm vessel was built by Gladding-Hearn of Somerset, Massachusetts. Constructed of welded aluminum, contained within the *Kane's* deep-V planing hull were two 400 hp GM diesel pumping engines and two 650 hp GM diesel propulsion engines. Operating from new quarters at the Brooklyn Navy Yard, the 52-foot *Kane* became the busiest fireboat in the world. The reserve *Smith* and the tender *Smoke II* also called the Navy Yard's Pier G home, as did Marine Division Headquarters and the Marine Repair Shops.

The *John J. Harvey* lay derelict at the Brooklyn Navy Yard until early 1999, when the city sold the boat in a sealed-bid auction. Her new owner had a tug move the fireboat to his pier on Manhattan's west side, where restoration efforts got underway. The fire department passed along the equipment it had stripped from the *Harvey*, which eventually was restored to working order by owners and volunteers. In 2000 it was placed on the National Register of Historic Places, but complete retirement just wasn't in the cards for the seventy-year-old *Harvey. It* still had life in it, but no one could possibly have imagined the extent to which it would be tested on September 11, the day the towers fell.

That morning, three of its owners and a couple of volunteers rushed to dockside, threw off *Harvey's* mooring lines, and headed south to the financial district to assist the Coast Guard with evacuating people. En route, the makeshift crew watched in horror as the towers fell. At Battery Park City they loaded about 150 people, which they took to Pier 40 a mile north. Fire officials then radioed to inform them that all water sources around the World Trade Center had been knocked out, and they needed the *Harvey* to do what it had done so well for decades—pump water to the fire.

So significant are the fireboats of New York City that they are paid homage by these scale versions displayed at the New York-New York Las Vegas Hotel and Casino. Author's Collection.

For three days, around the clock, the *Harvey* was again working alongside sister fireboats, as though it had never left. It became an island, offering brief respite for firefighters and others who "worked the pile."

As quickly as Marine Co. 1, the *McKean,* arrived and tied up at the seawall, people trapped at Battery Park jumped in desperation 15 feet to 20 feet down onto its deck to escape the blinding, choking cloud resulting from the collapse. As ladders were raised, some panicked people had to be stopped from undoing mooring lines. After taking a boatload of evacuees across the river to New Jersey, Marine Co. 1 returned to begin pumping operations. The *McKean* would not leave the scene until two weeks later, when water service was restored. The crew of the *Fire Fighter* also sprang into action at the sight of the burning North Tower of the World Trade Center. Arriving at the Battery, it pumped at capacity for three days and spent two additional days doing intermittent pumping, supplying water for operations at Ground Zero. The tender *Smoke II* joined the effort, carrying Fire Captain

The never-say-die *Harvey* not only outlasted every FDNY fireboat, it was called upon to pump water at Ground Zero on 9/11. The *Harvey* often is considered New York's first modern fireboat, having set the standard for generations of large fireboats. The powerboat mounted amidships was a later addition equipped with a 150-gpm pump for under-pier fires. Courtesy of John A. Calderone.

Marine 2, the *John J. Harvey*, puts on an impressive display, its streams framing the Statue of Liberty. Courtesy of Ron Frehm.

Al Fuentes of the Marine Division from the site to a New Jersey trauma center after he was rescued from the rubble. It returned to begin pumping operations. The *Kane* was laid up for repairs; however, its crew quickly placed it back in service and rushed to the scene. The *Kane* also evacuated people from lower Manhattan and then supplied water. For three days, the fireboats pumped about 60,000 gallons of water per minute to fire apparatus, building standpipe systems, and through manifolds. That they were the only water source available during the disaster would profoundly affect the design of New York's future fireboats.

According to a list compiled by John Doswell for Fireboat.org, more than 160 vessels were involved in firefighting, evacuation, and relief efforts during the 9/11 disaster. Working alongside FDNY boats *Fire Fighter*, *McKean*, *Kane*, and *Smoke II,* along with the *Harvey*, were vessels of the US Coast Guard, Merchant Marine, NYPD, tugs, ferry boats, and sightseeing and cruise boats.

Afterward, the *Fire Fighter* was inspected and overhauled before resuming duty at its Homeport station. Although it was deemed fit for more years of service, it was obvious that changes in marine operations brought about by technology and terrorism required extensive modifications. Given the boat's age and the price tag that went with revamping, its days were numbered. Nevertheless, for nearly a decade *Fire Fighter* maintained vigil over the Verrazano Narrows and routinely answered alarms. It fought its last major fire aboard a vessel when a gasoline barge exploded at Port Mobile on February 21, 2003. Later that year it played a major role in the Staten Island Ferry Disaster, and was on scene after US Airways Flight 1549 landed on the Hudson in January 2009.

In 1996, the FDNY Marine Division acquired a pair of former US Navy 27-foot Boston Whaler Vigilants. This Vigilant was assigned on a seasonal basis to Marine 3 at Beach Channel. Courtesy of John A. Calderone.

The fireboat tender *Smoke II* heads for the "deep six" to begin its new career as an artificial reef in the Sea Girt Reef off the coast of New Jersey. Its retirement in July 2008 marked the end of more than a century of tenders of the FDNY fleet. Author's Collection.

From jumbo fireboats to this utility-rescue boat, personnel of the FDNY's Marine Division skillfully made use of its equipment to maintain a high level of proficiency. Courtesy of John A. Calderone.

In the decade following 9/11, as FDNY officials reassessed the fleet and future needs, their mantra became "fast, strong, and agile." They agreed that their vessels had to be a combination of fast and light boats for water rescue, and large, strong boats to manage major responses and address the national security threat. The department adopted a tiered response system to commit its marine resources as a situation dictated. The uppermost tier comprises large, ultra-modern fireboats capable of handing the worst imaginable disasters. Like modern warships, they can serve as safe command platforms in chemical, biological, radiological, and nuclear (CBRN) environments. The rapid response tier comprises small, fast, fire-rescue boats. The tiered response at the operational level involves hundreds of firefighters trained to work aboard fireboats, operate flood rescue boats, and perform surf and ice rescues. Increased reliance has been placed on the use of foam concentrate and Purple K dry chemical agent for faster extinguishment and to avoid flooding vessels. The extreme pumping capability through monitors of battleship-

When Captain Chesley B. Sullenberger successfully ditched US Airways Flight 1549 on the Hudson River on January 15, 2009, the FDNY dispatched four vessels. Marine 1 Alpha is seen next to the Airbus, while North Hudson's Marine 1, at right, closes the distance. Marine 1, the fireboat *McKean*, follows. North Hudson Regional Fire Department covers municipalities north of Jersey City and Hoboken. Courtesy of Ron Jeffers.

The FDNY Marine Division's jumbo fireboat, Marine 1, the *Three Forty Three,* was named in honor of the 343 firefighters killed in the 9/11 attacks. The mighty 50,000-gpm fireboat began service on September 11, 2010. While under way, it can pump 25,000 gpm. Courtesy of John A. Calderone.

class fireboats can be used to dissipate chemical clouds and break up oil spills.

All FDNY fireboat pilots hold a US Coast Guard Harbor Pilot's license, while marine engineers are licensed at the Coast Guard's engineer level. Firefighters assigned to the boats complete mandatory training through the marine division.

Since the *Fire Fighter* had been earmarked for retirement since 9/11, early in 2006 plans for its replacement were released. Proof of the continuing need for battleship-class fireboats was evident in the plans, which called for a pair of 140-foot, conventional-hull fireboats, each capable of pumping 50,000 gpm.

Addressing the demand for modern, smaller fire-rescue boats in the tiered response system began in 2006 when a 27-foot quick-response boat was assigned to Marine Co. 1 to replace a Boston Whaler that had been assigned to the company in 1999. Boston Whalers had been acquired during the 1990s for distribution to marine companies; two had gone to the *Fire Fighter*, along with an Avon rubber inflatable boat. Marine Co. 1's small boat was built by Safe Boats International with a rigid foam hull, making it virtually unsinkable. Twin 250 hp outboards allow *Marine 1 Alpha* to skim across the water at 50 mph, making it ideal for responses to people in the water and small boats in distress. On board is a small fire pump and EMS equipment.

Over the course of the decade, other small boats for seasonal utility work and fire rescue were added to the inventory, most of which were trucked across the country from Safe Boats of Bremerton, Washington. Founded in 1997, the firm relies on its signature wrap-around foam collar, which ensures flotation, stability, and fendering. The same foam, which can be fortified with anti-ballistic material, is used to insulate the Alaskan oil pipeline.

The year 2008 signaled the end of the fleet's century-long use of fireboat tenders. The fifty-year old *Smoke II* was retired on July 15 while serving as backup for Marine Co. 6, the *Kane. It* was towed from the Brooklyn Navy Yard to begin a new career as an artificial reef off the coast of New Jersey.

The plans initiated in 2006 for a pair of jumbo fireboats led to an extensive design phase, which included assistance of US Navy engineers of the Joint Program Executive Office for Chemical and Biological Defense and Naval Sea Systems Command. The contract for the vessels was let in 2007 to Eastern Shipbuilding Group of Panama City, Florida. Eastern founder and owner Brian D'Isernia opened the yard in 1976 and soon specialized in a broad range of custom vessels, large and small. Eastern quickly gained a reputation as an innovative leader in the industry, and up to the challenge of building any type of vessel.

The first of the two fireboats was appropriately named the *Three Forty Three* to memorialize the 343 New York firefighters killed on 9/11. It was completed in late 2009 and after six months of training was put into service on September 11, 2010. When Fire Commissioner Salvatore Cassano, along with New

This view of the *Three Forty Three* is a study in complexity, showing the vessel's intricate details. Under wraps on a launch ramp is a 17-foot fast boat with 90 hp outboard. The boat is removed when connection is made to a 12-inch-diameter discharge outlet for pumping to high-capacity flow units such as Iron Man, Terminator, or Neptune. Courtesy of Glenn Duda.

York City Mayor Michael Bloomberg, commissioned the boat, he stated:

> THE NAME OF THIS SHIP WILL REMIND EVERY PERSON WHO SEES IT IN ACTION PROTECTING THE PORT OF NEW YORK AND SURROUNDING WATERWAYS, AND EVERY FIREFIGHTER WHO OPERATES ABOARD IT, OF THE SUPREME SACRIFICE MADE BY SO MANY OF OUR MEMBERS ON 9/11, ENSURING THAT WE NEVER FORGET (WWW.NYC.GOV, NOVEMBER 17, 2010).

Her sister ship, named *Fire Fighter II*, after undergoing sea trials, arrived in November, ending the seventy-two-year career of the *Fire Fighter*. The old warhorse served on reserve status alongside the also-retired *John D. McKean* at the Brooklyn Navy Yard until October 2012, when it was transferred to the Fireboat Fire Fighter Museum.

The twin jumbo fireboats were designed by the noted naval architectural firm of Robert Allan Ltd. of Vancouver, B.C. Much of their $54 million price tag was funded by grants from the Department of Homeland Security. Not only do they pump the highest volume of water of any fireboat in the world, they are the most technically advanced. Weighing 500 tons and measuring 140 feet, with a 36-foot beam and 16-foot draft, each ship relies on four 2,245 hp diesel powerplants built by MTU America for propulsion and pumping. Use of all four engines for pumping at standstill delivers 50,000 gpm, and 25,000 gpm are delivered using two engines while underway. Twelve monitors by Fire Fighting Systems (FFS) are remotely controlled, one of which is rated at an impressive 17,000 gpm, while six can deliver 6,000 gpm, and four at 2,000 gpm. Four foam pumps are also installed. Four variable-pitch screws and skegs allow a top speed of 21 mph and superior maneuverability. A bow thruster aids in fixed positioning and maneuvering.

A crucial element of the boats is a pressurized, filtered-air command post that can accommodate forty people. Normal crew complement is seven, along with twenty-seven firefighters if needed. Facilities for decontamination and medical treatment are also

The builder's plate of the fireboat *Three Forty Three*. Traditionally, such markers were mounted to the exterior face of the pilothouse. Courtesy of Glenn Duda.

Sister to the *Three Forty Three* is the *Fire Fighter II*, with steel hull and triple-aluminum superstructure, topped by a pilothouse giving all-around visibility. Its Allied System 3,000-pound capacity crane incorporates a telescoping boom-bucket equipped with a 1,500-gpm monitor. Courtesy of John A. Calderone.

built in. One of its many innovative designs is a forward ballast tank that can be filled in eight minutes for lowering the bow to match deck levels of large ferries that ply the waterways. Another feature is a hydraulic tower ladder/crane for ship deck operations. The *Three Forty Three* began service as Marine Co. 1 at Pier 40 on Manhattan's west side. The *Fire Fighter II* operates as Marine Co. 9 from new quarters that opened its doors in spring 2012. The station, along with a protective basin, was erected at the site of the *Fire Fighter's* berth at Homeport Pier in the Stapleton section of Staten Island.

Filling the gap between the twin jumbo fireboats and the fleet of small, fast boats is the 65-foot *Bravest,* built by Safe Boats. Commissioned on May 26, 2011, at Manhattan's North Cove Marina, the *Bravest* replaced Marine Co. 6, the *Kevin C. Kane*, which joined reserve boats at the navy yard. Although it displaces 40 tons, it can travel 50 mph thanks to triple Caterpillar C-18 1,000 hp diesel engines driving triple Hamilton waterjets. One of its two pumps is driven from the main engine, which uncouples from a main waterjet, while two outer waterjets provide propulsion for maneuvering. With shallow water capability, *Bravest* is a do-anything, go-anywhere vessel. Its 7,000-gpm rating, which supplies four monitors, is enhanced by 200-gallon foam and 100-pound Purple K systems. Like the big boys, its cabin has CBRN protection and a full electronic suite. Its nameplate is made from steel recovered from the World Trade Center.

Other small boats are continually added to the fleet, such as 33-foot, 600 hp Special Purpose EMT boats, and Marine Co. 8's 33-foot seasonal boat commissioned in 2012, all built by Safe Boats.

In this age of extreme, fast-paced technology, at least two of the veteran battleship-class fireboats of the FDNY are still around. After much needless politicking and environmental controversy, the *Fire Fighter*, the longest-serving New York fireboat, finally

This view of the *Three Forty Three* presenting a water salute captures a rare formation of the entire NYPD helicopter fleet. Environmentally safe vegetable dyes are injected through pickup tubes to generate colored streams. Courtesy of John M. Calderone.

found a home at Greenport Harbor, New York, where it undergoes restoration by admirers and volunteers. It got a new lease on life in 2015 when the Fireboat Museum was awarded a national Maritime Heritage Grant. The *John J. Harvey*, now in private ownership and operation, and also in the care of the restoration-minded, is moored at Pier 66 Maritime on the Hudson River, never far from where it did its best work for more than sixty years.

Overshadowed by New York's fireboat fleet, but no less important, are fireboats on duty throughout the state of New York. Yonkers, in Westchester County, lies along the Hudson River and borders the Bronx. The Yonkers Fire Department Marine Operations Unit operates the *James J. Mulcahey* as *Fireboat 1*, a former USCG 41-footer, to cover more than four miles of one of the nation's busiest waterways.

North of Yonkers is the Tappan Zee Bridge, which opened in 1956 to span the Hudson as a vital link in the New York State Thruway. The Bridge Maintenance Department operated a crash truck, two pumpers, and the 40-foot fireboat *Jay Bee II*. Besides protecting the bridge area with a 600-gpm fire pump, foam, and rescue equipment, the boat assisted municipalities along the riverfront.

Marine 6, Brooklyn's Bravest, features Safe Boats's signature wrap-around foam collar, which incorporates a push knee, along with a tow device for pushing or pulling vessels in distress. All top surfaces are of non-skid material. Courtesy of John A. Calderone.

Marine 1 Alpha, which firefighters often called The Small Boat, was delivered from Safe Boats in April 2006. Its basic construction is aluminum with rigid foam collar. Twin 250 hp outboards propel the 27-foot boat to 50 mph. Courtesy of John A. Calderone.

The fireboat *Abram S. Hewitt* in its heyday. On August 13, 1913, at a four-alarm fire at Standard Oil Company, fireman Brereton Johnson was killed aboard the *Hewitt* when a monitor tore loose. Courtesy of George F. Mand Library, FDNY.

The fireboat *Abram S. Hewitt*, after retirement in 1958, joined other once-proud vessels whose careers ended at the marine graveyard Witte Marine Scrapyard, Staten Island. Courtesy of John Landers-Beth Klein Collection.

Northport Fire-Rescue on Long Island operates a Model M2-33 delivered by Moose Boats in 2006. This is the first commercial outboard the company built since it opened its doors in 2000. Besides firefighting and rescue, the all-aluminum catamaran is fully EMS-equipped, earning it certification as a New York State ambulance.

The Yonkers Fire Department Marine Operations Unit covers more than four miles of one of the nation's busiest waterways, the Hudson River and Yonkers waterfront. Yonkers is immediately north of New York City in Westchester County. The department's Fireboat 1, the *James J. Mulcahey*, is a former Coast Guard 41-foot boat powered by twin Cummins diesel engines, with a 375-gpm pump. Courtesy of John A. Calderone.

For close-in work, the Yonkers Fire Department operates this fiberglass Shamrock with 175-gpm pump. Courtesy of John A. Calderone.

Northpoint Fire Rescue, Long Island, New York, operates this 500 hp catamaran from Moose Boats. The Stang bow monitor is fed from a 500-gpm Hale pump. A night vision infrared camera, along with communications and navigation gear, is incorporated into a folding mast. The vessel also serves as a New York State–certified ambulance Class C fireboat. Courtesy of Moose Boats, Inc.

Newark

The centuries-old city of Newark, the largest in New Jersey, had organized fire protection as early as the late eighteenth century. As Newark Harbor evolved into a busy port, in 1921 the Newark Fire Department purchased its first fireboat, the Newarker, formerly a Buffalo police patrol boat launched in 1885 and named Grover Cleveland. The 63-footer was equipped with two Ahrens Fox pumps. Updated fire protection for Port Newark would not occur until post-WWII when the city acquired a pair of Wheeler boats from military surplus. The ex-USCG 40-footers were named *William J. Brennan*, after New Jersey's Supreme Court justice, and *Michael P. Duffy*, honoring the city commissioner.

The Newark Fire Department got in step with the switch to lime-green for fire apparatus during the 1970s. The firm Ward LaFrance was marketing the color for its fire trucks, so Newark's order for pumpers triggered the changeover in 1973. The fireboat *Kennedy* was painted accordingly. The unpopular scheme remained into the 1990s, when the switch was made to white, and then to everyone's relief, back to traditional red. Courtesy of Curt Elie.

Newark, New Jersey's, fireboat *Newarker* is seen berthed at Point-No-Point, which is a railroad bridge crossing the Passaic River between Newark and Kearny, New Jersey. Author's Collection.

Especially interesting was their role in 1947 as participants in what the media called "The Battle of Newark Bay." With postwar plans to develop the port, City of Newark administration wanted the navy to give up its wartime base at the port. Navy officials side-stepped the simmering issue when a salvage firm that had purchased the battleship USS *New Mexico* headed it from Boston to Newark to be scrapped. Determined to prevent use of Port Newark as a scrapyard for warships, the fireboats *Brennan* and *Duffy* were ordered to prepare to fend off the incoming tow with their monitors and join tugs in forming a blockade. When the equally determined salvage firm added tugs to its effort, the stage was set for a most unusual naval battle. Fortunately, the Undersecretary of the Navy intervened and arranged a settlement that allowed a limited time frame for the salvage company to scrap three warships, and guarantee that no permanent scrapyard would be established at Newark. In an effort to smooth over discord with New Mexico, which took offense to the slur towards its namesake veteran warship, the ships were greeted with great fanfare, including a water salute by the two fireboats initially ordered to block their passage.

Newark Fire Department officials in the 1960s began a search for a new fireboat that not only combined the gpm output of the two Wheeler boats, but possessed additional features. Since Port Newark was fast becoming the largest cargo facility in the New York and New Jersey harbor region, miles of docks and warehouses had to be quickly reached by a maneuverable vessel that could navigate safely in shoal waters.

Their search ended at Fireboat Division, Gladding-Hearn Shipbuilding Co. at Somerset, Massachusetts, which was no stranger to crafting fireboats. During construction of hull number 86, the bow of the welded steel hull was given extra-thick plating to protect it from ice. In the engine room were four 300 hp engines, two of which drove twin screws for a top speed of 15 mph, while two operated a pair of 2,000-gpm Peerless pumps. The boat was named *John F. Kennedy* and measured 48 feet long, with a 15-foot beam. Drafting only 3 feet for work in very shallow water, its screws and rudder were protected from damage in the shallows by "grounding shoes" welded beneath them.

Three McIntyre 1,100-gpm monitors were deck-mounted, with the foredeck monitor adaptable for a National Foam distributor that drew from a 200-gallon

The City of Newark placed this 3,000 SeaArk Guardian in service in December 2006. Later, the 48-foot fireboat wore different markings and was designated Fireboat 1. Courtesy of SeaArk Marine.

supply of liquid concentrate. Other features of the *Kennedy* included a fixed carbon-dioxide system to protect fuel and engine compartments, 2,000 feet of hose, and a 6,000-watt generator.

Replacing the *Kennedy* in December 2006 was a 48-foot, 3,000-gpm SeaArk first named *Guardian,* and then *Fireboat 1.* Power is derived from a pair of 800 hp Caterpillar engines. An internal 100-gallon foam concentrate tank gives the crew an edge in fighting class B (flammable liquid) fires. Like most modern vessels acquired with Homeland Security funding, Newark's fireboat has provisions for CBRN monitoring.

During 2012 and 2013, eight jurisdictions in northern New Jersey took delivery of 28-foot, 1,000-gpm boats for fire rescue. All are Boston Whaler Vigilants purchased with Homeland Security funding. Modeled after the FDNY Marine Division's Vigilant, the boats are powered by twin 300 hp outboard engines, giving them a top speed of 50 mph. All are equipped with intra-service compatible radios and infrared radar.

Since Jersey City, New Jersey, is bordered on the east and west by busy waterways, the Jersey City Fire Department in 2006 acquired from SeaArk a twin to Newark's fireboat. SeaArk, which began in 1959 as MonArk Boat Company in Monticello, Arkansas, has remained a family-owned business that has survived due to its wide range of fire-rescue and special-purpose boats. Jersey City Fire Department's vessel was christened *The Joseph Lovero* on September 29, 2006. The name was an easy choice because of Joe Lovero's dedication to the people and the city of Jersey City. He had served as an EMT, was a life-long member of the Gong Club, and worked as a fire department dispatcher. Three days after his sixtieth birthday, on September 11, 2001, Lovero rushed to the World Trade Center to help and was killed in the collapse.

Home Security Port funding for northern New Jersey went beyond fireboat acquisitions. Working in conjunction with the boats are Marine Coastal Response Units (MCRU) that consist of land apparatus staffed by firefighters who not only crew the boats, but are trained in other specialties. These units fall under the Urban Area Security Initiative (UASI), which was created as a regional response effort that shares personnel and equipment. Members receive training from the Coast Guard and as hazardous material and CBRNE technicians. The initiative categorizes the Jersey City–Newark area as one of the ten highest risk regions in the nation.

Jersey City's fireboat *Joseph Lovero* was a twin to Newark's SeaArk fireboat delivered in 2006. Joseph Lovero, a member of the fire service and the Gong Club, was killed in the collapse of the World Trade center on 9/11. Courtesy of the Gong Club.

Jersey City's *Lovero* and FDNY's *Governor Alfred E. Smith* share dock space in 2007 after the *Lovero's* commissioning, where the Smith gave a water display. Blue barrels aboard the *Smith* are filled with foam concentrate. Courtesy of Connie Spellman.

Besides the fireboat *Joseph Lovero*, the Jersey City Fire Department Marine Coastal Response Unit operates Marine 2, a 28-foot Boston Whaler, along with four Zodiac boats. Courtesy of John A. Calderone.

This Firematic Brush Rapid Attack Truck (BRAT) was acquired for the Marine Coastal Response Unit of the Jersey City Fire Department and stationed near the fireboat *Lovero* with a crew of four. Built on a 2014 Ford F550 high-water chassis, the BRAT features a 250-gpm diesel pump and 6-ton winch. Courtesy of Ron Jeffers.

On the western edge of New Jersey, bordered by the Delaware River, is the city of Camden, whose fire department operates the fireboat *George M. Jackson*. The vessel is named in honor of a Camden firefighter who died from injuries sustained in a fire in 1990. Designers and builders at MetalCraft Marine pulled out all the stops to fit a powerful package within the *Jackson's* 50-foot hull. Triple 750 hp diesel engines drive triple Hamilton jets for a top speed of 49 mph, and power triple Hale pumps to deliver 11,000 gpm. It is equipped with full electronic suites and provisions for CBRNE operations. The Firestorm 50 is a local and regional multi-mission craft able to operate fifty miles offshore, primarily on the Delaware River. While performing that mission, the *Jackson* is in good company, for across the Delaware are the fireboats of Philadelphia.

Camden, New Jersey, fire-rescue's Marine 1 is a former USCG 32-foot Ports and Waterways Boat acquired in 1994. The 500-gpm vessel was named the *William Ables* after Camden's first fire chief. Courtesy of Dale Mutza.

The borough Edgewater, New Jersey, along the Hudson River, was one of eight municipalities in 2012 and 2013 to receive a 28-foot 1,000-gpm Boston Whaler Vigilant under Homeland Security funding. Others were North Hudson, Newark, Jersey City, Linden, Carteret, Hoboken, and Elizabeth. Courtesy of Ron Jeffers.

Replacing Camden's William Ables in December 2011 was the *George M. Jackson*, named in honor of a Camden firefighter killed in the line of duty. The MetalCraft Marine 50-foot vessel uses waterjet propulsion and is capable of pumping 11,000 gpm. The lower section of the push knee slides upward to reveal a monitor for low-angle work. Courtesy of Phil Cohen.

North Hudson Fire-Rescue, New Jersey, puts its 3,800-gpm Firestorm 36 through its paces in 2014. Courtesy of Ron Jeffers.

Twin 490 hp Cummins diesels power twin water jets of Perth Amboy's Marine 5. The 2,500-gpm Firestorm 36 began service during 2005. Courtesy of MetalCraft Marine, Inc.

This high-capacity flow unit, named Iron Man, was built by National Foam-Kiddie with a 12 inch diameter intake supplying an 8-inch-diameter monitor capable of delivering 8,000 gpm at 115 psi. Although weighing 3 tons, an integral ballast water tank and jack stands are necessary to keep the unit stable. Here, Iron Man of the New Jersey's Union County Neptune Task Force is supplied 8,000 gpm from the FDNY fireboat *Fire Fighter II* to battle a five-alarm blaze at Keasbey, New Jersey, on April 25, 2014. Courtesy of John P. Melfa.

During the 1950s, the Atlantic City, New Jersey, fire department used this military surplus DUKW, commonly called Duck, as a fireboat, complete with pump, monitor, and firefighting equipment. Courtesy of Bob Bartosz.

Philadelphia

City officials recognized early in Philadelphia's history the need for fireboats, joining Boston and New York in acquiring vessels for waterfront protection. Long before a pure fireboat was built for the Bureau of Fire, however, police boats equipped with fire pumps doubled as fireboats along Philadelphia's Delaware and Schuylkill Rivers. In keeping with its reputation for innovation, Philadelphia is also credited with creating the first large-scale, high-pressure water main system for firefighting, and establishment of a Bureau of Ice Boats. Unlike other large cities that relied on their fireboats to keep ice-choked waterways open, Philadelphia operated a fleet of massive, 637-gross-ton, steam-powered paddlewheel ice boats.

The police boats, named after city officials and mayors of Philadelphia, were assigned to the Harbor Patrols of the Delaware and Schuylkill Rivers. They varied in size and included the *Samuel H. Ashbridge, Visitor*, *John E. Reyburn*, *Samuel G. King*, and *William S. Stokley.* The latter reportedly was the largest at 96 feet long and capable of pumping 1,000 gpm. The keel was laid for the *Stokley* in 1874, four years after city fathers established a professional, paid fire department. It went in service in 1975, and was assisted by the *King,* which entered service in 1881. It was built as hull number 218 by the experienced shipbuilder John H. Dialogue, owner of River Iron Works across the river in Camden. Weighing 63 gross tons, the *King* measured 80 feet long and delivered

The 110-foot fireboat *Edwin S. Stuart* at Philadelphia's historic Race Street Pier in 1908. White was a popular color for fireboats of the period. The appearance of a single monitor on a large fireboat was misleading since pumps more often were used to supply numerous nozzles manned along bulwarks. Courtesy of Jack Wright Collection.

One of the nation's largest fireboats built during the nineteenth century was Philadelphia's first firefighting vessel, the *Edwin S. Stuart.* Typical of fireboats of the era, long deckhouses accommodate large crews and emphasized the boat's sweeping sheer. Courtesy of Jack Wright Collection.

1,200 gpm.

To keep pace with the growth of Philadelphia's port, the first dedicated fireboat was built for the Bureau of Fire. Named after Mayor Edwin Sydney Stuart, who lobbied for the boat in 1892, the *Edwin S. Stuart* arrived in 1893 and began service as *Fireboat 1* on September 1 at Pier 10 North Wharves, Race Street on the Delaware River. Initially equipped with only one monitor, the *Stuart* was rated at 3,000 gpm until the installation of two Clapp & Jones pumps boosted the rating to 4,500 gpm; turn-of-the-century annual reports show gpm outputs as 6,000 gpm and even 9,000 gpm. The steel-hulled, 105-foot *Stuart* was built by Cowles Engineering Company of Brooklyn, New York.

By 1918, the Harbor Police fleet had been reduced to three small boats, and the *Stuart* was deemed inadequate to go it alone protecting the many miles of developing river shoreline. That prompted design proposals for a pair of similar fireboats, the contracts for which were signed with two shipbuilders. In 1921, J. W. Sullivan Co. of New York City began work on the *Rudolph Blankenburg,* while Merchant Marine on the southern edge of Philadelphia in Chester began building the *J. Hampton Moore* as hull number 387.

Plans called for a length of 130 feet for the *Blankenburg* and 121 feet for the *Moore;* both had 28-foot beams. The *Moore* drafted 9 feet and displaced 292 tons, compared to the *Blankenburg's* nearly 13-foot draft and 350-ton displacement. Four Davidson pumps in each boat delivered 10,000 gpm, and both vessels featured three deck monitors and a 28-foot tower, plus two 15-foot towers on the same deck.

The *J. Hampton Moore* at Pier 56 South in July 1938. Monitors of the *Moore*, and its sister ship the *Blankenburg*, were mounted high for maximum stream reach. Three bands painted near the top of its forward funnel differentiated it from the *Blankenburg*, which wore two bands. Courtesy of David M. Boone Collection.

Philadelphia's *Rudolph Blankenburg* moored at Pier 10 North Wharves, which incorporated fireboat quarters. After reconstruction during the 1930s, the pier was re-numbered Pier 11 North Wharves. Courtesy of David M. Boone Collection.

The *Moore,* which was designed by Philadelphia naval architects Cornell and Matthews, was powered by a 750 hp vertical compound engine, the steam for which was supplied by two Babcock & Wilcox oil-burning boilers. Although a monitor atop the pilothouse and those of the towers were rated at 3,000 gpm at 250 pounds pressure each, the *Moore* struggled through pre-acceptance tests, showing 10,000 gpm, then considered somewhat disappointing for its size.

The *Blankenburg* was launched first, on August 9, 1921, at the New Jersey Dock Co., christened by the late mayor's widow, Lucretia Blankenburg.

The *Moore* slid down the ways at Merchant Marine on December 20, 1921, christened by Mrs. W. C. K. Walls. Both fireboats began service in 1922, the *Moore* in February.

Philadelphia's three fireboats served admirably for more than a quarter century, when it was decided to replace the antiquated trio. Extensive planning went into the first fireboat, which, although considered the first large fireboat built after WWII, would not have the large dimensions familiar to early vessels. City planners theorized that multiple, smaller fireboats were better than one large boat, echoing noted fireboat designer William Cowles.

The *Bernard Samuel* was the first of the trio built by RTC Shipbuilding within a three-year period for the Philadelphia Fire Department. Immediately abaft the aft monitor is a towing bitt. The *Samuel*'s trim tanks could take on water to add heft to the boat for towing operations. Courtesy of Jack Wright Collection.

First among the new trio of fireboats was the 76-foot *Bernard Samuel,* designed by Naval Architect Thomas D. Bowes and built by RTC Shipbuilding, across the river at Camden. Displacing 94 tons, the *Samuel* incorporated numerous design features that promised strength, versatility, and maneuverability. Its stout, well-rounded steel hull, combined with a large rudder and single screw, enabled the boat to be driven on and off ice and mud flats. Fore and aft trim tanks also allowed the *Samuel* to work in shallows by lessening the boat's draft. Electric and hand steering, inherent stability, and simple machinery allowed ease of operation by a small crew.

Fireboat 1, the *Delaware*, in its element in 1963, when it was berthed at Pier 109 South Wharves on the Delaware River. Later, it would be repainted red overall, including its tower and hose reel covers. Author's Collection.

The *Samuel's* versatility resulted mainly from its 800 hp powerplant. The General Motors Series Quad-6 comprised four six-cylinder diesel engines that could simultaneously power the boat's drive system and four 1,500-gpm De Laval two-stage centrifugal pumps. Its 6,000-gpm punch could be delivered through three McIntyre Brass Works 1,500-gpm monitors, one of which was atop a 24-foot elevating tower built by Michigan Tool Company. Filling out the *Samuel's* firefighting complement were six hydrants and two massive, deck-mounted Wirt & Knox swivel hose reels. In later years, the boat would carry more than 1,000 gallons of foam concentrate.

After acceptance trials, during which the boat reached speeds approaching 18 mph, the *Samuel* entered service in November 1948. The *Stuart* was retired and scrapped the following year.

The elaborate planning that went into the *Samuel* paved the way for the next two fireboats, which would be even more alike than the pair they were to replace. Like the *Samuel,* the sister ships, named *Benjamin Franklin* and *Delaware,* were designed by Thomas Bowes and built by RTC Shipbuilding. Both vessels featured improvements over the *Bernard Samuel.* Diesel-powered, with a top speed of 17 mph, each had four 1,500-gpm De Laval pumps to deliver 6,000

In profile, the *Delaware* displays its practical arrangement of monitors, tower, and hose reels. The monitor tower, although it reached 24 feet above the waterline, could be used at any angle. Author's Collection.

gpm through deck monitors and one atop a 24-foot hinged tower. Both boats measured 80 feet, with a 19-foot beam, and drafted 7½ feet. A 500-gallon foam tank in each boat later was increased to 1,000 gallons.

The *Franklin* (hull number 198) went into service first, on May 25, 1950, just one week after delivery. Its twin, *Delaware,* (hull number 199) arrived in Philadelphia on June 23 and began service on July 1, 1950. With their arrival, the *Blankenburg*, Fireboat 2; and the *Moore*, Fireboat 3; were removed from front line service.

In November 1972, Philadelphia's three fireboats were reorganized as engine companies: the *Delaware*, then berthed at Race Street Pier 11 on the Delaware River, became Engine 15; the *Samuel* at Allegheny Avenue on the Delaware became Engine 23, and the *Franklin*, on the Schuylkill River became Engine 32. Two years later they were renamed Marine Companies.

The *Samuel* was relegated to reserve status in 1983 until its disposal in 2008.

Land apparatus of the Philadelphia Fire Department that could be used in conjunction with fireboats included this 1949 Reo Giant Deluge Wagon, seen here working at a six-alarm warehouse blaze in August 1992. The monitor of *Giant Deluge* was rated at 2,000 gpm. The Reo truck is a descendant of apparatus SS 99 (Special Services), when the department had five such units in service. Courtesy of Frank Saia/Curt Elie Collection

Philadelphia Fire Department's *Bernard Samuel* pumps to produce thirteen streams using its monitors and multiple portable nozzles. Courtesy of Jack Wright Collection.

In January 1975, Marine 32, the *Franklin*, responded from its berth on the east bank of the Schuylkill River near the Passyunk Bridge to a tanker collision at a refinery south of Philadelphia. Marine 32's officer ordered the boat against the pier so that trapped firefighters could jump aboard to safety. Courtesy of Dennis Shell Collection.

Alongside the base of a tower of the *Benjamin Franklin Bridge*, the Samuel's crew connects to a standpipe so that water can be pumped up to the bridge roadway. The monitor tower is in its fully lowered position. Courtesy of Jack Wright Collection.

The *Delaware*, seen in the third color scheme for fireboats of the Philadelphia Fire Department. The *Delaware* was re-designated Marine Unit 1 in July 1993. Courtesy of Jack Wright.

Federal firefighters of the US Navy-Philadelphia Naval Shipyard operate this former US Coast Guard 32-foot vessel, which is rated at 500 gpm. Courtesy of John M. Calderone.

The modern, 5,500-gpm fireboat *Independence* puts on a water display along Philadelphia's waterfront in October 2010. Courtesy of Patricia Calderone.

Built by Derecktor Connecticut as hull number 86040, Philadelphia's fireboat *Independence* weighs 65 gross tons. After arrival in 2007, the vessel took up station at the US Coast Guard Station on the Delaware River. Courtesy of Jack Wright.

During 2007, the Philadelphia Fire Department enlarged its fireboat fleet with the *Independence*, built by Direcktor Shipyards, Bridgeport, Connecticut. Rated at 5,500 gpm, the all-aluminum fireboat was placed in service in December. It measures 66 feet, having a 20-foot beam and 2-foot, 8-inch draft. It can attain a speed of 32 mph.

Philadelphia's Marine Division is governed by a Special Operations Command, with *Marine Unit 1* berthed at the US Coast Guard station on the Delaware River and *Marine Unit 2* on the Schuylkill River, plus reserve marinc units.

Dclaware

Wilmington, Delaware's largest city, lies less than twenty miles south of Philadelphia along the busy Delaware River. For more than a quarter century the Wilmington Fire Department Marine Unit has provided fireboat protection locally and in the tri-state area, working closely with law enforcement and Philadelphia-based Coast Guard units.

To replace the outdated fireboat acquired from Baltimore County, in 1989 the fire department acquired a 65-foot utility vessel built in 1977 as hull number EP-209 by Blount Marine Corporation, Warren, Rhode Island. Conversion to a 7,200-gpm fireboat included a Dorchester pump installed in 1990 along with seven monitors. Twin 350 hp Detroit diesel engines gave the 80-gross-ton vessel a top speed of 15 mph. In keeping with thc department practice of assigning engine company numbers to correspond with station numbers, the fireboat was designated Fireboat 7; numbers 1 through 6 designate land engine companies. The vessel was later named *Firefighter.* The fireboat, along with smaller vessels designated Marine 8 and Marine 9, was stationed on the Christiana River, which flows into the Delaware.

To keep pace with rapid expansion along the Christiana riverfront, Fireboat 7 was replaced in 2012 by a Firestorm 70 built by MetalCraft Marine. Funded

The Wilmington, Delaware, fire department in 1989 acquired this twelve-year-old utility vessel and converted it to a fireboat that could pump more than 7,000 gpm. Fircboat 7 served until 2012, when it was sold. Courtesy of Curt Elie.

Still flying the flag of Canada, its country of origin, Wilmington, Delaware's new Fireboat 7 arrives in July 2012, flanked by a Delaware State Police boat. The Firestorm 48 is capable of pumping 12,000 gpm. Courtesy of MetalCraft Marine, Inc.

mainly by a federal port security grant, the new Fireboat 7, *Firefighter*, arrived in July and was christened on September 19. Although nearly identical in dimensions to its predecessor, it was significantly lighter, displacing 52 tons. Four 825 hp diesel engines power four Hamilton jets for a top speed of 44 mph and superior maneuverability. Its 12,000-gpm firepower is delivered by four Hale pumps through five monitors, one of which is a 5,000-gpm remote-controlled Stang roof monitor. Other features include 300 gallons of foam concentrate, bow thruster, and hydraulic crane. A push knee on the bow and CBRNE capability have become standard features on fireboats produced by MetalCraft Marine.

Farther downriver is the town of New Castle, home to Holloway Terrace Fire Co., which operates a 2008 27-foot Boston Whaler Challenger with 75 hp Darley pump. Next on a southward track is Delaware

Holloway Terrace Fire Company of New Castle County, Delaware, operates this 2008 27-foot Boston Whaler Challenger powered by dual 300 hp outboard engines. Courtesy of Louis P. Angeli.

Remote-control roof and bow monitors, plus a manually operated monitor aft, provide knockdown power for this Firestorm 30 of the Delaware City Fire Company. Fireboat 15 is a 2011 MetalCraft Marine with jet drive that propels the boat to 45 mph. The vessel sank after crashing into a rock jetty in June 2014. It was raised, repaired, and resumed service in 2015. Courtesy of MetalCraft Marine.

Marine 89 is a 27-foot SeaArk Commanders VC operated by Memorial Volunteer Fire Company, Slaughter Beach, Delaware. Built in 2006, the vessel features an extended cabin and foam fendering system. Courtesy of SeaArk Marine, Inc.

City's 2011 35-foot, 1,750-gpm fireboat. Farther down the Delaware coast in Delaware Bay is Slaughter Beach, where Memorial Volunteer Fire Company operates a 27-foot SeaArk Commander acquired in 2006. Designated Marine 89, the fire-rescue boat features a foam fendering system and extended cabin. Where Delaware Bay meets the Atlantic Ocean, the Lewes Fire Department operates the largest fire-rescue boat in the lower Delaware Bay region. Moose Boats of Petaluma, California, delivered the vessel in December 2008. Following a rigorous training program, 82-Marine 1 began all-weather regional coverage in May 2009. Moose Boats specializes in aluminum, water-jet-driven catamarans for public service and the military. The Model M1-44 measures 44 feet, with a 16-foot beam and 2-foot draft. Twin Cummins 600 hp turbo diesels power Hamilton waterjets and a Waterous pump delivering 900 gpm through two monitors.

The Lewes fire-rescue boat is equipped to operate as a multi-agency command post during major incidents, while protecting waterways that see heavy cargo and passenger travel. Annually, more than 70 million tons of cargo transit Delaware Bay, more than 10 percent of which is imported crude oil headed for six refineries along the Delaware River.

Delaware's Lewes Fire-Rescue operates M1-44 from Moose Boats. Not only does the boat pump 900 gpm, it is equipped to function as a regional response craft and command post during major incidents. All decks are of non-skid material, including an extended stern dive platform. Courtesy of Moose Boats, Inc.

Baltimore

Two decades would pass between Baltimore Fire Chief John W. Watkins' request in 1870 for a fireboat and the city's purchase of a vessel. In 1890, budget appropriations were approved for a fireboat and crew, and a contract let to E. J. Codd Co. Designated Engine Co. No. 16 and named *Cataract*, the steam-powered, 85-foot, 4,400-gpm fireboat was placed in service on July 1, 1891, at Commercial Wharf in Baltimore's Fells Point. When the Great Baltimore Fire swept through the city's commercial and waterfront district on February 7, 1904, the *Cataract*, assisted by three tugs and the revenue cutter *William Windom*, was credited with halting the conflagration.

In September 1909, Engine Co. No. 16 relocated to larger accommodations at the Inner Harbor, where a second fireboat was to be assigned. Designated Engine Co. No. 39 and named *Deluge,* the new fireboat joined the *Cataract* during March 1911. Measuring 120 feet long and weighing 315 tons, the steam-powered *Deluge* could pump 9,000 gpm.

The Marine Division of the Baltimore Fire Department was established on April 22, 1912, and the following year the *Cataract* went to Baltimore's Spedden Shipbuilding for extensive overhaul. Reappearing in 1914 as a newly built vessel, the 185-ton *Cataract* now measured 89 feet and could pump 7,000 gpm.

Baltimore's first fireboat, the 120-foot *Cataract*, assumed protection of the waterfront in 1891. Seen here, on her afterdeck during the late 1930s, are unique collapsible life rafts with flat pods. Author's Collection.

In 1917, the *Deluge* was relocated to Fort McHenry to provide better coverage of the waterfront, which in two years had increased from fifteen miles to forty-five miles of shoreline due to industrialization and annexation. Such expansion called for two additional

Baltimore's 330-ton fireboat *Deluge* works at a lumberyard blaze in August 1937. The 120-foot, 9,000-gpm *Deluge* joined Baltimore's first fireboat, the *Cataract*, in 1911. Tarpaulins over trunk deck areas were common for relief from summer sun during long duration at alarms. Author's Collection.

When the commanding officer of Baltimore Fire Department's Marine Division was designated Battalion Chief No. 12 in early 1922, he was assigned this tender. The 60-foot, steam-powered vessel came from the army where it served as the *John C. Gregg.* The fire department stationed the tender with Engine Company No. 16 at Pier 7, President Street. Author's Collection.

fireboats, one of which was a submarine chaser (S.C.428) loaned to the city by the US Navy on May 16, 1921. Built by Mathis Yacht Building Co. of Camden, New Jersey, as hull number 97, the wood-hulled sub chaser had been commissioned on December 28, 1918, and decommissioned in March 1921. A slender boat at 110 feet with a 13-foot beam, it was powered by three Hall Scott 220 hp gasoline engines. WWI planners had called for a small sub hunter so that large numbers could be built in civilian shipyards, thereby freeing major yards for large warships. The vessel's 3-inch gun, machine guns, and depth charge launcher were replaced by deck monitors and nozzle mounts. A fire pump gave it a rating of 2,800 gpm. After outfitting, the vessel was designated Engine Co. No. 48 and named *Cascade;* it was assigned to the industrial Curtis Bay area in June 1921. With a top speed of 20 mph, the *Cascade* was the fastest of the fleet.

In 1922, the battalion chief commanding the marine division was assigned a 60-foot, 300-gpm tender, which came from the army where it served as the *John C. Gregg.* That same year the largest fireboat in Baltimore's history arrived from Baltimore Dry Docks. The all-steel vessel was christened *Torrent* by Ethel Broening, daughter of Mayor William F. Broening, and designated Engine Co. No. 48. The 340-ton steam-powered vessel stretched 121 feet, with a 29-foot beam and a 13-foot draft. With the ability to pump 9,000 gpm, the *Torrent* was the obvious choice for berthing at the industrial district of Canton at Baltimore's outer harbor. It was converted from coal to oil in 1935, and when converted to diesel power the following year, its pump rating was boosted

The ex-US Navy sub chaser *Cascade* joined the Baltimore Fire Department after conversion to fireboat in 1921. Installation of a fire pump allowed it to pump nearly 3,000 gpm. It was designated Engine Company No. 49 and assigned to Baltimore's industrial Curtis Bay area. Author's Collection.

to 11,000 gpm.

Baltimore depended on its four fireboats for more than three decades until the results of an inspection and survey of the aging fleet proved that the *Torrent* was the most costly in terms of repair and operating costs. So with the arrival of a new all-steel fireboat, the *Torrent* was sold to a local shipyard to embark on a new career as a tug. In June 1949, the city of Baltimore assumed ownership of the *Cascade*.

Baltimore's fireboat Engine Company No. 48, the *Torrent*, after "pleasure-boating" in February 1936. Despite year-round service through harsh winters, the *Torrent* served for more than three decades, when it began a new career as a tug. Author's Collection.

Aft monitors of the *Torrent* are tested during a test of the new two-watt radio system in April 1938. Long ladders with hooks used for boarding ships were attached to the aft tower. Author's Collection.

The largest fireboat in the history of the Baltimore Fire Department was the *Torrent*, seen here transiting Baltimore Harbor during the 1930s, prior to being converted from steam power to diesel. Author's Collection.

When it was nearly a quarter of a century old, Baltimore's *Cataract* emerged from rebuild in 1914 a new fireboat capable of pumping 7,000 gpm. Although constructed of steel, its main deck was wood. Hooked ladders were fastened to the monitor tower. The *Cataract* served as Engine Company No. 16 at Marine Division Headquarters, Pier 7, at the foot of President Street, finally retiring in 1960. Courtesy of Curt Elie Collection.

Named in honor of Thomas D'Alesandro Jr., Baltimore's mayor from 1947 to 1959, the fireboat was designed by marine architect Thomas D. Bowers of Philadelphia and built by RTC Shipbuilding of Camden as hull number 234. Weighing 149 gross tons, the vessel relied on a pair of 660 hp Fairbanks-Morse diesel engines to drive the 103-foot fireboat at its top speed of 17 mph. A second pair of engines powered fire pumps that provided 12,000 gpm. Noteworthy was the boat's ability to maintain full propulsion power and pump at capacity simultaneously. Nicknamed "Tommy," the *Mayor Thomas D'Alesandro, Jr.* was placed in service on 16 September 1956.

Since the *Cataract, Deluge*, and *Cascade* were becoming obsolete, plans were made for replacing all three to fully modernize the fleet. Approval was granted for the purchase of three identical fireboats, with construction begun at Jakobson Shipyard, Oyster Bay, New York.

During construction of the trio, the tender too was replaced, thanks to the benevolence of the Baltimore Police Department. Among its fleet was a wooden 37-foot cruiser, which had been launched on April 26, 1956, as a police patrol boat. Built by Edwards Boat Co., Middle River, Maryland, as an Edwards Trunk Cabin Cruiser, it was named *Beverly Ober* after the Baltimore police commissioner from 1949 to 1955. In 1959, the police department gave the boat to the fire department marine division, where

The large sign atop the building at Municipal Pier 7 read "Fireboat Station Cataract." The site served as Marine Division Headquarters and home to Engine Companies 39 and 16. The *Cataract* is seen here in its black and red scheme during the 1950s. Author's Collection.

Baltimore Fire Department's second tender was the *Beverly Ober*, turned over by the Baltimore Police Department in 1959. Its twin screws were driven by two gasoline-powered 125 hp Chrysler marine engines. Aboard for a trial cruise were, from left to right, Fire Board president James Lacy, fire department chief John T. Killian, police commissioner James Hepbron, and fire department secretary John O'Malley. William Klender/Courtesy of Curt Elie Collection.

The first of four fireboats to usher out the Baltimore City Fire Department's fleet during the late 1950s was the powerful, all-steel *Mayor Thomas D'Alesandro*. The *"Tommy's"* four monitors were included in the white portion of this striking color scheme. Courtesy of Howard Meile III.

it would serve until it was sold in 1971.

All three new fireboats were 85-foot vessels weighing 93 tons. Propulsion was a single screw powered by a Fairbanks-Morse diesel engine, with Fairbanks-Morse centrifugal fire pumps providing 6,000 gpm at 150 psi. Hull number 397 was named *Mayor J. Harold Grady,* hull number 398 the *P. W. Wilkinson,* and hull number 399 the *August Emrich.* Pinkney Wilkinson was executive secretary of the Baltimore Fire Department, and Emrich was chief engineer during the city's 1904 Great Fire. When the trio began service in October 1960, the *Cascade* was returned to the navy, having served thirty-nine years as a Baltimore fireboat.

When industrialization caused significant downturn in activity along Baltimore's waterfront during the early 1970s, Engine Co. No. 49 at Wagner's Point was disbanded in December 1971, followed by the closing of Engine Co. No. 48 with the *Torrent* at Colgate Creek in June 1972. Reorganization of the marine division then had Engine Co. No. 16 labeled Fireboat No.1, and Engine Co. No. 39 re-designated Fireboat No. 2. The Marine Battalion was disbanded in March 1974, with the State of Maryland assuming some financial responsibility for port fire protection.

As industry along the northwest portion of Baltimore's harbor all but disappeared, expansive marina development necessitated acquisition of a fast response boat in 1988. Equipped with a pump and bow monitor, the two-man vessel could skim along on 8 inches of water and get into tight spots inaccessible to the big boats. Crewed by two firefighter/EMTs, Fire-Rescue 1 took up station with Fireboat 2, the *Grady,* at Fort McHenry.

In December 1991, Fireboat 1, along with four

Two of the *Deluge*'s powerful streams from monitors are being used to break up ice between Baltimore's many piers in February 1945. Courtesy of Curt Elie Collection.

land engine companies, fell victim to the budget axe, leaving Fireboat 2 and the fire-rescue boat in service; the *D'Alesandro* became Fireboat 1 on 1 July 1992. The reserve *Wilkinson* and *Emrich* were decommissioned in 1994, with the former sold to Vane Brothers Company as an oil spill response vessel.

Early in the twenty-first century, Baltimore officials contracted with SeaArk Marine of Monticello, Arkansas, for a pair of 30-foot Enforcer models, each capable of pumping 1,500 gpm. Both are able to reach speeds of 40 mph and are equipped for firefighting and medical emergencies. With arrival of the pair in 2003, a 1,000-gpm Boston Whaler acquired in 1991 was sold to the Riviera Beach Volunteer Fire Department,

Baltimore Fire Department's first small boat was this 24-foot fast response vessel stationed with a crew of two at Fort McHenry. Since it was also responsible for emergency medical services, firefighter/EMTs aboard could treat a patient for transport by a medic unit waiting at a predetermined location. Author's Collection.

Fire-Rescue 2 was one of two SeaArk Model 3010 Enforcers that arrived in Baltimore in 2003. Both are capable of pumping 1,500 gpm and are equipped for medical emergencies. Courtesy of Howard Meile III.

The 85-foot *August Emrich* when it began service during late 1960, along with two identical fireboats. Courtesy of John Floyd II, Royal Blue Ltd. Archives.

During the 1970s, Baltimore County Fire Department operated two boats to cover its 173 miles of shoreline along Chesapeake Bay and its tributaries. The custom-built W. H. Wideholt was purchased in 1978, but proved problematic and never went in service. It was sold to Wilmington, Delaware, and sank at the Delaware River Pier in February 1982. Courtesy of Curt Elie Collection.

Maryland.

In keeping with the nationwide trend for regional response watercraft that swept the nation following 9/11, Baltimore City Fire Department took advantage of Department of Homeland Security funds to acquire a modern fireboat. The timing could not have been better since the fireboat *Grady* was approaching the half-century mark. That Fireboat 1 and Fire-Rescue 1 were averaging 100 calls per year in Baltimore's sprawling harbor and along miles of Patapsco River shoreline guaranteed approval of a new fireboat.

The contract was let to Ontario's Hike Metal Products Ltd. for an 87-foot steel-hull fireboat capable of pumping 13,000 gpm at 150 psi. Twin Caterpillar

Baltimore City Fire Department fireboats wore this striking high-visibility scheme until the beginning of the twenty-first century, when they were re-painted traditional white over red. The 85-foot, 6,000-gpm *Grady* was built as hull number 397 at Jakobson Shipyard, Oyster Bay, New York. Reserve 2, the *D'Alesandro*, is in the background. Courtesy of Howard Meile III.

The largest fireboat on Chesapeake Bay is Baltimore's *John R. Frazier*, see here shortly after delivery in 2007. The 7,000-gpm *Frazier* also serves as a CBRNE Regional Response Vessel. Courtesy of John A. Calderone.

1,650 hp diesel engines drive twin screws, giving the 157-ton boat a top speed of 18 mph. A pair of single-stage, centrifugal fire pumps are each driven by a Caterpillar 715 hp diesel engine. Water, or foam from two 500-gallon concentrate tanks, is delivered through four monitors, two of which are remote-controlled, along with sixteen discharge outlets.

The vessel left the firm's Lake Erie yard on June 2, 2007, arriving one week later at Marine Division Headquarters on Leahy Street near Fort McHenry. On August 4 it was christened *John R. Frazier* in honor of Baltimore's Fire Prevention Bureau commander and Maryland EMS advocate who died earlier that year. With a crew of four—pilot, officer, marine engineer, and second engineer or firefighter—Fireboat No. 1, *Frazier*, is the only fully staffed regional response vessel, serving Anne Arundel, Baltimore, Carroll, Harford and Howard counties, and Annapolis. In addition, it is the only large fireboat on the Chesapeake Bay. The *D'Alesandro* became reserve Fireboat 2. Fire-Rescue Boat No. 1 operates from the same station, while Fire-Rescue Boat No. 2 is kept in reserve. The fireboat *Grady* was sold in 2008. In 2015, the *D'Alesandro* succumbed to budget woes when imminent repairs proved too costly, and it was removed from service.

Monitors, hose reels, and discharge manifolds made for a crowded deck of the *Mayor J. Harold Grady*, seen here in its final color scheme. Author's Collection.

Besides Baltimore, more than half of Maryland's twenty-three counties front Chesapeake Bay or the Atlantic Ocean, necessitating fire-rescue vessels. A number of fire departments representing these counties are equipped with MetalCraft Marine fire-rescue boats purchased during the twenty-first century.

Apparatus of the Chesapeake Beach Volunteer Fire & Rescue Department of Virginia Beach, Virginia, were painted white, including this military surplus DUKW. A unique feature of the amphibian was a crow's nest with floodlights. Courtesy of Martin Grube.

During September 2010, Moose Boats delivered this 35-foot aluminum catamaran boat to Anne Arundel County, Maryland. Named the *Arundel Patriot*, the vessel is operated jointly by the Anne Arundel County Fire and Police Departments for emergencies in Chesapeake Bay and Annapolis. A fire pump powered by a 5-liter Mercruiser engine delivers 1,000 gpm. Propulsion power is derived from twin 300 hp outboards. Courtesy of Moose Boats.

Throughout its history, the Riviera Beach Volunteer Fire Company, Maryland, operated a wide variety of fire-rescue boats. Although the manufacturer of this WWII-era boat is unknown, it bears strong resemblance to 45-foot Air Rescue Boats built during the war by Peterson Shipbuilding of Sturgeon Bay, Wisconsin. Here, modified with a fire pump and elevated bow and stern monitors during the 1950s, it held its own against fires. Courtesy of John Floyd II, Royal Blue Ltd. Archives.

Formerly of the Baltimore City Fire Department, this 28-foot Boston Whaler is one of four tasked with responding to emergencies beyond Chesapeake's 533 miles of shoreline. Maryland's Riviera Beach Volunteer Fire Company serves the waterfront community of Chesapeake Bay, south of Baltimore City. Courtesy of Curt Elie Collection.

In late 2006, the Annapolis, Maryland, fire department took delivery of this Firestorm 36. The boat was purchased with a Homeland Security grant to protect twenty-six miles of waterfront. The vessel replaced a smaller fireboat, which had been built at Station 35 by department personnel. Courtesy of MetalCraft Marine, Inc.

Anne Arundel County's 43-foot MetalCraft Marine fire-rescue boat is one of four such vessels that provide protective services for Chesapeake Bay. Reverse-angle windshields, which are standard on MetalCraft Marine vessels, not only aid vision in poor weather, they reduce glare. Courtesy of Curt Elie.

Federal firefighters of the Aberdeen Proving Grounds Fire Department use Marine 12 on the waterways of the US Army base in Maryland's Harford County. In 2014, the department took delivery of a 27-foot North River Almar Sounder fireboat rated at 500 gpm. Courtesy of John M. Calderone.

This 36-foot, 1,750-gpm fireboat, operated by Prince George County, became the fourth emergency vessel on the Potomac River. Courtesy of MetalCraft Marine, Inc.

Washington, DC

Fire protection of the waterways bordering the nation's capital was begun immediately following the Civil War with the US Navy tug USS *Rescue*. The steam-powered vessel displaced 111 tons and measured 80 feet long with an 18-foot beam. Built by Harlan and Hollingsworth of Wilmington, Delaware, in 1861, the vessel served as a gunboat during the war with the Union Navy's Potomac Flotilla. It was converted to a fire tug in Philadelphia and home-ported at Washington Navy Yard to augment then–City of Washington's newly formed Paid Steam Fire Department. The department became fully paid on September 23, 1871, becoming the District of Columbia Fire Department. The *Rescue* served as the city's fireboat until 1889.

The District of Columbia took delivery of its first purpose-built fireboat in June 1905. Weighing 192 gross tons, the boat left Harlan and Hollingsworth's yard as hull number 354 and was placed in service on December 8 as Engine Co. 19. The 101-foot vessel bore the name *Firefighter* on its hull, which housed Duplex pumps delivering 6,000 gpm. In September 1920, its company assignment was changed to Fireboat No. 1 when land company Engine 19 was established. Providing assistance to the *Firefighter* beginning in 1929 was the 53-foot patrol-fireboat *Charles A. Evans* built as hull number 264 at Spedden Shipbuilding.

The *Firefighter* remained in service for forty-one years, when it was replaced by another fireboat of nearly equal size and power. The 110-foot large harbor tug *USS Maquinna* (YTB-225) was assigned to the Potomac River Naval Command during WWII. Displacing 410 tons, the diesel-powered tug was launched at the yard of John Trumpy & Sons, Gloucester, New Jersey, in April 1944. In late 1946 the navy transferred the tug to the District of Columbia Fire Department for use as a fireboat. After outfitting, painted red, and named *William T. Belt,* the fireboat served until 1959; it's anyone's guess where it ended up after that.

The large harbor tug USS *Maquinna* was only two years old when the navy transferred the powerful 110-foot tug to the District of Columbia Fire Department for use as a fireboat. It was named the *William T. Belt* and served until 1959. Courtesy of Dick Nailen.

The nation's capital enjoyed fire protection on the waterways beginning at the end of the Civil War. This 101-foot fireboat, named *Firefighter*, arrived in 1905 and served for forty-one years before being replaced. The *Firefighter* could deliver 6,000 gpm to its six monitors. Courtesy of Curt Elie.

When the FDNY sold its fireboat *John H. Glenn Jr.* to the District of Columbia Fire Department in 1977, *it* was fifteen years old, relatively young for a 70-foot fireboat. The *Glenn* was designated Fireboat No. 1, and in 1981 it was joined at the old Coast Guard wharf by Fireboat No. 2, a former USCG patrol boat with 225-gpm pump. With plenty of power for propulsion and pumping, the *Glenn* did the job. It took a tragedy, however, to highlight a design deficiency. When Air Florida Flight 90 crashed into the Potomac River in January 1982, the *Glenn's* crew encountered difficulty negotiating the ice-choked river to rescue survivors. Consequently, in 1984, the *Glenn's* bow was reinforced to 14-inch thickness to maneuver in heavy ice conditions.

For faster response on the Potomac and Anacostia Rivers, in 1996 a 25-foot Boston Whaler Guardian was placed in service as Fire-Rescue 3. After complete refurbishment of the *Glenn* in 2004, which included replacement of its three engines, it could pump 7,000 gpm. Although thought to have many good years left to serve after the two-year long overhaul, the *John Glenn* did not fare as well as its astronaut-senator namesake. In January 2009 a cruise boat slammed

When the fireboat *John H. Glenn Jr.* went to the District of Columbia Fire Department to become Fireboat 1, it kept its name. Here, in 1983, the *Glenn* and Fireboat 2, a former Coast Guard patrol boat, are berthed at the Fireboat Station, Pier 5. Courtesy of Jim Senese.

Throughout its years protecting the waterfront of the nation's capital, the fireboat *John H. Glenn Jr.* persevered through refurbishment and repair, while retaining its status as Fireboat 1. It began life with the FDNY in 1962. Author's Collection.

into it at dockside, tearing a 15-foot long gash in the bow. The damage, along with numerous leaks, placed it out of service for an extended period. In 2012 the Inspector General's office declared that the boat was obsolete and incapable of managing a major waterfront disaster, including a terror attack, on the nation's capital. But the half-century-old *Glenn* persevered and in March 2006 was joined by a new Fireboat 2, replacing the Boston Whaler. The Firestorm 30 from MetalCraft Marine's long line of fire-rescue boats was a welcome addition to the DC Fire Department. Purchased with Homeland Security funds, Fireboat 2 is rated at 1,800 gpm and can travel at speeds topping 40 mph.

MetalCraft Marine credits the Occoquan Woodbridge Lorton, or OWL, Volunteer Fire Department of Woodbridge, Virginia, with providing valuable input in the development of its fireboats. Members of the department, twenty miles south of Washington, DC, at Occoquan Harbor of the Potomac, helped design the Firestorm 36, which was delivered in 2002.

Between Woodbridge and Washington, DC, is Alexandria, Virginia, a community rich in seafaring tradition, although having relinquished its status as a prominent seaport. After the City of Alexandria Fire Department in 1997 formed a Marine Operations Team, it followed through with its intent to acquire a modern fireboat to replace an outdated vessel. With the aid of a Port Security grant, Fireboat 201, named *Vigilant*, was acquired in 2003. The 30-foot, 1,500-gpm fire-rescue boat was equipped to meet the requirements of *NFPA Directive 1925* as a class C fireboat. Assisting the *Vigilant* were two Zodiac boats, which were towed by fire apparatus.

Realizing the need for a larger, more powerful vessel, Alexandria city officials in 2010 proposed that the *Vigilant* be replaced by a 44-foot, 4,000-gpm vessel. A revised requirement for additional water flow meant a larger vessel, since gpm flow is related to vessel size. Replacing the *Vigilant* at City Marina in September 2013 is a Firestorm 50, rated at 7,400 gpm and named *Relief*. Like its predecessor, it would wear number 201 since its crew was assigned to Engine Co. 201.

Joining the fireboat *Glenn* in 2006 was this Firestorm 30 from Canada's MetalCraft Marine, having replaced a Boston Whaler to increase waterfront protection. The 1,800-gpm vessel mirrors the markings worn by the department's land apparatus. Courtesy of MetalCraft Marine, Inc.

Streams from Washington, DC's, Fireboat 2 frame the nation's capital. Courtesy of MetalCraft Marine, Inc.

Fireboat 6 is one of three vessels supporting the Marine Team of the Virginia Beach Fire Department. This 24-foot Grady White Shark served as a police boat prior to joining the fire department in 2007. Twin outboards drive the boat at 52 mph, with fire suppression capability provided by a 500-gpm pump. Courtesy of Virginia Beach Fire Department.

The third boat acquired in 2007 for the Virginia Beach Fire Department Marine Team is this 30-foot Argus Class Northwind, rated at 1,500 gpm. Courtesy of Virginia Beach Fire Department.

North of Alexandria, Fairfax County Fire and Rescue Department operates a Firestorm 36 named *Gunston Hall.* Elsewhere along Virginia's Atlantic coast, the Marine Team of the Virginia Beach Fire Department operates three vessels, designated Fireboats 1, 6, and 12. Fireboat 12 is largest among them, a 41-foot former Coast Guard utility boat, with another in reserve. Members of the Marine Team assigned to Company 1 are also members of the regional Marine Incident Response Team, which comprises area fire departments.

The *Vigilant* of the fire department of Alexandria, Virginia, displaying the lines of fireboats built by MetalCraft Marine, is captured here making a high-speed run when it was new in 2003. A decade later, it would be replaced by an even larger fireboat named *Relief*, also from the yard of MetalCraft Marine. Courtesy of MetalCraft Marine, Inc.

The Fairfax, Virginia, County Fire and Rescue Department christened this Firestorm 36 *Gunston Hall* on June 2, 2005. Gunston Hall is a historic Virginia landmark, home of US founding father George Mason. Twin Cummins diesels drive the *Gunston Hall*'s waterjets, and give it a rating of 1,750 gpm. Courtesy of MetalCraft Marine.

Norfolk

Norfolk has the second largest population in Virginia, next to Virginia Beach, and is steeped in naval history. Bordered by multiple bodies of water and at the center of the Hampton Roads metro area, the city has an equally extensive history of fireboats. Fire protection for Norfolk's vast waterfront originally consisted of privately owned fireboats.

The Norfolk Fire Department's first experiment with a fire tug involved the steam vessel *Nonpareil* (SP-370) built in 1904 by Robert Palmer & Son Shipbuilding & Marine Railway Co. at Noank, Connecticut. After chartering the boat from a civilian firm, the navy took ownership in 1917, designating it District Harbor Tug YT-40. On November 29, 1920, the navy loaned the boat to Norfolk for use as Fireboat 29. Although re-powered with Sterling engines, it was returned to the navy after seven months. The Palmer Yard closed its doors in 1914 but was reactivated for WWII boat production as Groton Iron Works.

Norfolk's second experience with a military surplus vessel showed only slightly more promise. A commercial tug built in 1904 by A. C. Brown & Sons joined the navy in 1917 to become USS *Bouker* (SP-1275 and later YT-30). *It* was put up for sale but withdrawn in July 1921 and transferred to the Norfolk Fire Department in August. Finding the vessel unsuitable as a fireboat, Norfolk city fathers in June 1922 returned the *Bouker* to the navy at Portsmouth.

The Occoquan Woodbridge Lorton (OWL) Volunteer Fire Department of Virginia operates this Firestorm 36, which is rated at 2,250 gpm. Department members were instrumental in the fireboat's design, which aided MetalCraft Marine in future fireboat design. Courtesy of MetalCraft Marine, Inc.

On the third try, the Norfolk Fire Department had more luck with the army, in 1923 acquiring a tug named *General Harvey Brown*. Built in 1909 by the Fore River Shipyard at Quincy, Massachusetts, as hull number 163, it was one of ten such boats built of 98-foot length, with a 22-foot beam and weighing 163 gross tons. After outfitting as a fireboat by the Norfolk Fire Department, it began service in 1923 as the *Vulcan*. The *Vulcan* served Norfolk until January 1941, when it was sold in 1942 to work the James River as the tanker *Vivian Powell*.

The Norfolk, Virginia, fire department's early experience with fireboats included the *Vulcan*, seen here in 1930. It was built in 1909 as an army tug used in support of coastal artillery. Large cowl ventilators rising above the after house extended far below to the engine room. Author's Collection.

In 1942, Navy Department funds allocated for the purchase of civilian boats included eight for the Coast Guard to protect Norfolk's harbor. One, for example, was the private yacht *Hoi Toi* (originally named *Valia*), built in 1925 at East Greenwich, Rhode Island, under registry number 224737. Given USCG designation CG-72001-F, it and the other seven boats, after conversion to fireboat, was under the charge of the Coast Guard Captain of the Port for Norfolk and Portsmouth Harbors, 5th Naval District. The vessels were strategically located throughout the harbor, with their main station called Fireboat Station No. 4. To respond to alarms, they were tied into the Norfolk Fire Department's alarm system. The *Hoi Toi* began service at Norfolk in June 1942, serving only until May 1946 when it was transferred to the navy, which was switching to diesel-powered district harbor tugs.

With Naval Station Norfolk's growth into the largest naval installation in the nation, the city enjoyed ample fire protection along the waterfront. That would change at the end of the twentieth century when the navy began its post-Cold War drawdown, which had many agencies assuming operations

The *Hoi Toi*, built in 1925, was one of sixteen civilian vessels measuring 65 feet and longer converted to fireboats by the US Coast Guard. The gasoline-powered 70-footer was one of eight conversions assigned to Norfolk Harbor during WWII. Here, the vessel, along with sister fireboats, use rigid drafting hoses to supply deck-mounted 500-gpm pumps. Courtesy of US Coast Guard.

previously managed by the navy. To bolster remaining commercial and Coast Guard marine firefighting capability, after more than six decades, Norfolk got back into the fireboat business.

On September 5, 2008, Norfolk Fire-Rescue took delivery of a 30-foot Argus class fireboat, which was named *Vulcan II,* the name having been resurrected from the department's first fireboat. The all-aluminum boat was built by Northwind Marine of Seattle. Power is derived from twin outboards, giving it a top speed of 40 mph, while an inboard engine powers a pump to deliver 3,000 gpm, and jet drive for maneuvering near people in the water. The *Vulcan II* also carries the designation Fireboat 13 since it is quartered with Engine 13, Ladder 13, and Medic 13 in Ocean View District.

In 2008 Norfolk Fire-Rescue added to its marine capability with this 30-foot Argus vessel built by Northwind Marine. Fireboat 13 was named *Vulcan II* in honor of the department's first fireboat. Courtesy of Curt Elie Collection.

With the great-great grandchildren of Thomas Kevill, Norfolk's first fire chief, in attendance, on June 21, 2013, Norfolk Fireboat 1, the *Thomas Kevill,* was commissioned. Likened to a modern floating fire engine, the vessel is MetalCraft Marine's stretched version of its Firebrand 28, the highest-pumping NFPA Class C fireboat in North America. Twin 250 hp outboards give the 37-foot, all-aluminum boat a top speed of 46 mph, while a V-8 pumping engine powers a Darley pump to deliver 1,500 gpm through two monitors.

At the mouth of the James River, along its north shore, lies the port city of Newport News, whose significant fireboat history begins with Newport News Shipbuilding and Drydock Company. Until the sprawling facility was purchased by industrial giant Northrop Grumman in 2001, this was the largest privately owned shipyard in America. Like many industrial complexes, the shipyard operated a proprietary fire department, which in 1983 acquired a fireboat. The 45-foot, all-aluminum vessel was MonArk Boat Company's largest fireboat to date, with power derived from twin Detroit diesel engines and twin Hale pumps delivering 4,000 gpm. Staffed by a fire and security-trained crew, the fireboat conducts 24-hour patrol of the shipyard's shoreline.

Fireboat 1 of Virginia's Newport News Fire Department proudly bears the name of one of the city's legends, BMCM Carl M. Brashear. Master chief boatswain's Mate Brashear was the first African American to become a US Navy Master Diver. More than 750 horsepower strong, the Firestorm 30 was dedicated on October 28, 2007. Courtesy of US Navy.

In a 2002 exchange of fire equipment that drew sharp criticism from officials, the nineteen-year-old fireboat, along with a pair of dated fire trucks, went to the Newport News Fire Department, which turned over to the Northrop Grumman Fire Department two refurbished fire trucks to be maintained by Newport News mechanics. Much of the disdain over the deal came from the excessive time the vessel sat in dry dock for repair. Despite insistence by the firefighter's union that a newer and better vessel was a more sound investment, Fire Chief Kenneth Jones expressed gratitude for having more efficient marine capabilities and less reliance on Coast Guard vessels based at distant Portsmouth.

In 2007 the Newport News Fire Department took delivery of a 30-foot Firestorm to replace its two older boats. Powered by twin Cummins diesel engines paired with Hamilton jets, the all-aluminum vessel reaches 44 mph, and its Darley pump delivers 2,300 gpm.

This 45-foot SeaArk Protector was built for the Newport News Shipbuilding Fire Department as a fire and security patrol vessel. Twin Hale pumps each are driven by 240bhp Detroit diesel engines, giving the boat a rating of 4,000 gpm. The bow monitor is capable of delivering 2,000 gpm, while stern monitors each deliver 1,000 gpm. Courtesy of SeaArk Marine, Inc.

The Carolinas

Wilmington is prominent in the history of fireboats of North Carolina's coastal region. Located in the southern portion of the state, Wilmington's first fireboat was a wooden vessel built by city shops in 1907. Named *Atlantic*, it was powered by a 25 hp Globe gasoline engine propelling it to 8 mph. A Silsby third-size steam engine delivered more than 600 gpm. Two crewmen were on duty at all times. The *Atlantic* served until 1914, when replaced by the *Atlantic II*, a wooden vessel also powered by a 25 hp gasoline engine. A 125 hp gasoline pump drove an American LaFrance 1,400-gpm pump. It was replaced in 1949 by *Atlantic III*, which had been built in Brownsville, Texas, in 1943 as a patrol boat. Conversion of the 65-foot wooden vessel to fireboat by Wilmington city shops included two Hale pumps powered by 150

Third in a long line of fireboats named *Atlantic* of the Wilmington, North Carolina, fire department, this wooden 65-footer was acquired in 1949. A 230 hp Superior diesel provided propulsion and two 150 hp Mack gasoline engines drove two Hale pumps to supply two tower monitors. The *Atlantic III* was built as a patrol boat in 1943. Courtesy of Curt Elie Collection.

Built as a boarding tug during the Great Depression, the *Atlantic IV* in 1963 was converted to a fireboat resembling its predecessor of the Wilmington Fire Department. Courtesy of Curt Elie Collection.

hp Mack gasoline engines, for a rating of nearly 3,000 gpm. Two monitor towers were added, along with eight hose outlets. The *Atlantic III* was powered by a 230 hp Superior diesel engine.

Wilmington's resourcefulness in acquiring and outfitting fireboats, all named *Atlantic,* became especially apparent with *Atlantic III's* replacement in 1963. Looking back into this vessel's history reveals that it was born of necessity during the Great Depression. In 1932 Bath Iron Works of Massachusetts approached the government for projects to keep their workers employed. Government officials obliged, contracting Bath to build two vessels: a lighthouse tender named *Hickory,* and a "boarding tug" for the US Public Health Service. The tug would carry medical personnel to ships to treat ill crewmen. The inspection boat was named *T. B. McClintic* for surgeon Dr. Thomas B. McLintic, who discovered the cure for Rocky Mountain Spotted Fever.

Originally a boarding tug for the Public Health Service, the *T. B. McClintic* was restored to its original configuration following twenty-three years of service as Wilmington's fireboat. Author's Collection.

The 64-foot boat was a mainstay in the government fleet until 1958, after which it saw light duty as an ice-breaker. In 1963 the Wilmington Fire Department purchased the boat for $300 and converted it to a fireboat, naming it *Atlantic IV.* Weighing 85 tons, the steel-hull, diesel-powered vessel was rated at 2,500 gpm. It served Wilmington until it was sold in 1986. Eventually it was restored to an original health tug and placed on the National Register of Historic Places as the last known remaining quarantine boarding tug afloat and in operating condition.

A former 1960s US Navy fireboat carried on the *Atlantic* name after undergoing a two-year renovation. The Wilmington Fire Department unveiled *Atlantic V* in 1986. The vessel sank in 1999 and was replaced by a 1970s former Coast Guard 32-foot Port and Waterways Boat. It became Fireboat 2 in 2007 with the arrival of *Atlantic VI,* a MetalCraft Marine Firestorm 50 rated at 7,000 gpm. The 1,660-mile delivery voyage from the Ontario, Canada, factory to Wilmington was the company's longest water delivery to date. After a century of providing marine protection with hand-built and refurbished fireboats, the Wilmington Fire Department welcomed its first exclusively designed and built fireboat, direct from the factory.

Wilmington Fire Department's *Atlantic V* is seen here after refurbishment in 1986. It began life as a US Navy fireboat and was sunk in 1999. Courtesy of Curt Elie Collection.

Until closure of the Charleston Naval Shipyard during the 1990s, the Charleston, South Carolina area relied on US Navy tugs with firefighting capability for waterfront protection. The Charleston Fire Department operated only a 21-foot boat equipped with a 300-gpm pump. Charleston officials then set their sights on a 30-foot fireboat rated at 2,000 gpm, but ended up purchasing from MetalCraft Marine a 36-footer rated at 3,800 gpm. The boat arrived in November 2012 and was named *Louis Behrens,* after Charleston's longest term fire chief. Additional waterfront protection for the Charleston area is derived from North Charleston, which operates a 2,000 gpm "Firestorm 32," St. John's Fire District to the southeast, which operates a 1,200 gpm MetalCraft Marine boat, and the Mount Pleasant Fire Department, across the harbor from Charleston, which in 2014 acquired a 28-footer rated at 1,250 gpm. Mount Pleasant fire officials had submitted specifications for the boat to designer Response Marine, of Massachusetts, which passed along the plans for fabrication by Viking Welding of Kensington, New Hampshire.

North of Charleston, in the Myrtle Beach area, Horry County operates a 30-foot fireboat rated at 1,500 gpm. The boat was one of two originally built in 2001 by the Silver Ships yard at Mobile, Alabama, for Orange Beach, Alabama. Horry County purchased the boat in 2009.

During the twenty-first century, the various marine response capabilities of waterfront communities along North and South Carolina coasts have undergone continuous upgrades in fire-rescue boats. Noteworthy in Charleston waters is the offshore emergency response craft *Fort Ripley*. Designed by C. Raymond Hunt Associates and built by Gladding-Hearn Shipbuilding, the 64-foot multi-mission vessel complies with Coast Guard regulations requiring chemical and oil tankers to have contingency assets when within fifty miles of US trading ports. The Charleston Branch Pilots' Association formed Southeast Ocean Response Services to operate the *Fort Ripley*, which covers the area between Morehead City, North Carolina, and St. Augustine, Florida. Besides being on call to provide offshore firefighting, the sea-going boat serves as fireboat for Charleston Harbor and doubles as backup pilot boat and offshore supply vessel.

Unique to the *Fort Ripley* is the first commercial application in North America of Volvo's triple Inboard Propulsion System (IPS) drive, combined with Dynamic Positioning System (DPS) for hands-off positioning under GPS control. The IPS comprises three small steerable engine pods, each powered by 700 hp D13 diesel engines turning forward-facing, counter-rotating propellers. The advantage over jet drive is measured not only in fuel efficiency, but in speed and stability. Designers at C. Raymond Hunt Associates ensured that the boat runs as though on rails, with no slide or vibration when turning. The center engine can be disengaged to power a 3,500-gpm Hale pump that supplies two monitors, along with manifolds to supply hoses on shore or aboard ships. Commissioned on October 7, 2014, the *Fort Ripley* is sure to be a selling point for IPS drive, which has already gained wide acceptance in European maritime operations.

To comply with Coast Guard regulations governing tanker ships, the Charleston Branch Pilot's Association acquired this 64-foot offshore vessel named *Fort Ripley*. Versatility was key in designing the all-aluminum, deep-V-hull boat. Special features include hinged platform "wings" on both sides of the pilothouse, and all-around pan-nose bumper, flat push nose, and tire fenders. Author's Collection.

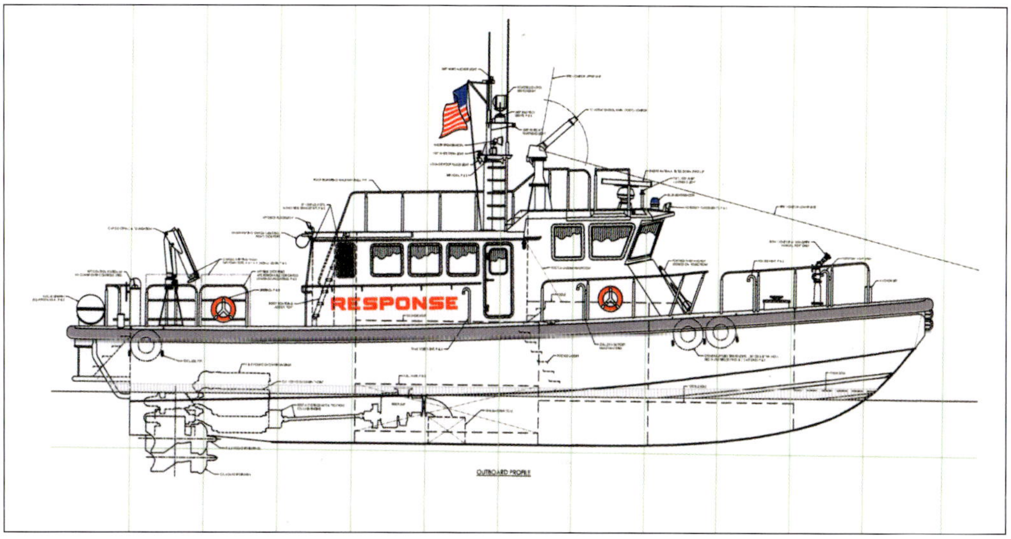

The outboard profile of Charleston's Response boat *Fort Ripley*. Author's Collection.

Certified by the Coast Guard to carry a twelve-person response team, the *Fort Ripley* also features a 1,000-pound-capacity crane, foam injection system, breathing apparatus charging system, and Pyrolance steel-penetrating nozzle system.

Georgia

Savannah's former fireboat is the largest among the various types of vessels that have protected Georgia's nearly 110 miles of coastline. At the northern tip of Georgia's coastline, historic Savannah records a long history as a major Atlantic seaport and industrial center.

Convinced that a fireboat was needed to fight shipboard fires, port officials and owners of private waterfront businesses during the 1980s began lobbying for an army tug that was slated for sinking as an artificial reef. Fire department officials sought the tug to replace its firefighting barge, which had to be pushed to a fire. The army tug was the *Dozier* (ST 483), which had been built during 1943 and 1944 by JK Welding of Brooklyn under design 257-A as hull number 76. Designs 257 and 257-A were the most popular for small army tugs, with eighty-eight vessels built. The *Dozier* and sister tugs began life with the army's Marine Design & Repair Branch of the Transportation Corps.

Savannah city officials were successful in acquiring the tug, which was passed from the Savannah District of Corps of Engineers to the fire department. After outfitting as a fireboat, which was funded by Savannah Port Authority, the boat arrived in January 1986 and began service in February. It was formally dedicated on March 24 as the *Edward J. Fogarty,* honoring the former head of Chatham County Civil Defense Council.

Throughout nearly six years of service, no fires occurred in the port area that required the *Fogarty's* services. Viewing the fireboat as nothing more than a drain on the city's budget, city officials threw caution to the wind and put the *Fogarty* up for sale. Delta Marine of Wilmington, North Carolina, submitted the winning bid with the intention of converting her back to tug configuration. Having worked on the boat in previous years, Delta Marine vice president W. R. Detyens bought it sight unseen, explaining, "I had a sentimental attachment to it." Back at work as a tug, the boat was re-named *Babette,* and later *Miss Peggy.* In 1996 it went to a dredging company in Canada, and at last report was in Charleston, South Carolina.

Savannah, Georgia's, fireboat *Edward J. Fogarty* began life as the army tug *Dozier* built during WWII. In keeping with the color scheme for land apparatus of the Savannah Fire Department, it was painted lime green and black. Author's Collection.

After decades without fireboat protection, in August 2015, Savannah, Georgia, took delivery of this 50-foot, 8,000-gpm MetalCraft marine "all-hazards" boat designated Marine 1 and christened *Courageous.* Courtesy of MetalCraft Marine, Inc.

Florida

Major cities along Florida's Atlantic coastline, Jacksonville, Fort Lauderdale, and Miami have a significant history of fireboats. Like a number of cities across America, Jacksonville suffered a disastrous conflagration around the turn of the century. The blaze began on May 3, 1901, destroying more than 2,000 buildings in its path and claiming seven lives. Although lessons were learned and changes were made to avert similar tragedies, some improvements took years to implement. The acquisition of fireboats often fell into that category due to their expense and operating costs. Jacksonville's fire chief Thomas Haney's prolonged lobbying for a fireboat during the turn of the century eventually resulted in the port city's first fireboat, the *John B. Callahan,* named in honor of a city councilman. The *Callahan* was a converted US Navy sub chaser built as hull number 97 by Vinyard Shipping Co., Milford, Delaware. The wooden, 110-foot vessel was commissioned March 13, 1918, at Philadelphia and designated SC-145. Just three years later it was sold into private ownership in Jacksonville, and on September 1, 1922, went into service with Jacksonville's fire department. The *Callahan* displaced 85 tons, had a beam of 13½ feet, and drafted 6 feet. Its four Sterling 220 hp gasoline engines gave it a top speed of 20 mph. With De Laval pumps, it was rated at 6,000 gpm. The *Callahan* would remain in service at its "temporary" quarters for forty-one years, retiring in 1963.

During its twenty-ninth year of duty, the *Callahan* would be joined by the fireboat *Richard D. Sutton,* which began service in Jacksonville on April 30, 1951. Replacing the *Callahan* in 1963 was Fireboat 82, a 63-footer rated at 4,000 gpm and capable of 20 mph. Built by Jacksonville's Gibb's Shipyard, No. 82 was delivered to the Jacksonville Fire Department on July 8. After two decades of service, the *Sutton* was decommissioned, with its pumps and monitors installed in its replacement. The *Sutton's* steel cabin ended up as a deck atop a Jacksonville building. Its replacement was a tug converted to fireboat by Jacksonville Shipyards in 1972. Named *Jax No. 1,* the 75-foot, 136-ton vessel pumped 2,000 gpm.

Modernization of the fleet came in 1971 with the 6,000-gpm *Eugene Johnson,* named for Lt. Newton E. Johnson, who had been killed in a fire the previous year. The 65-foot, all-aluminum vessel was built by Sewart Seacraft of Berwick, Louisiana.

The fireboat Marine One, which also carries the name of former Jacksonville mayor *Jake Godbold,* arrived in March 2007. Built by MetalCraft Marine, the *Godbold's* twin diesel engines provide 1,700 hp for jet propulsion, giving the boat a top speed of 50 mph, and a pumping capability of 6,500 gpm.

Jacksonville Fire and Rescue's Marine Unit, which operates out of two fire stations, also operates four smaller boats.

Fort Lauderdale's more than 100 miles of waterways and canals has earned it the title "The Venice of America." Fittingly, for more than half a century the city has operated a variety of fireboats. Fort Lauderdale's first fireboat on record was a 28-foot, steel-welded

The fireboat fleet of the fire department of Jacksonville, Florida, was first considered modernized when it took delivery of this 65-foot, all-aluminum boat named *Eugene Johnson.* Built in 1971, the *Johnson* was powered by three Detroit diesel engines, with three Fairbanks-Morse pumps, each rated at 2,000 gpm. Courtesy of Curt Elie.

The ultimate in modernization for Jacksonville fireboats, Marine 1, the *Jake Godbold*, arrived in 2007. Here, the *Godbold* pumps water through its five monitors in 2010. Twin diesel engines drive the vessel's water jets for a top speed of 50 mph. Courtesy of MetalCraft Marine.

SaftiCraft built in 1960. The craft was introduced in 1950 by Avondale Marine Ways of Westwego, Louisiana. Avondale was acquired in 1951 by A. P. and E. W. Dont of Morgan City, Louisiana, which continued production of SaftiCraft models. Titled Fireboat 1 and nicknamed "Old Reliable," the boat was used cooperatively by Fort Lauderdale fire and police departments. The boat could deliver 1,500 gpm through two Mystery deck monitors. In addition, it mounted five discharge outlets and carried dewatering eductors to bail vessels. Old Reliable drafted 3 feet and could reach speeds approaching 40 mph.

Records indicate that Fort Lauderdale purchased a second fireboat in 1964, which was a 26-foot, steel-hull boat with jet drive. It served until 1970 when replaced by a fireboat converted from a 37-foot Striker sport-fishing boat. The boat was built in 1969, reportedly at the company's Bergen, Norway, yard. The Striker Yacht Corp. opened its doors in 1951 at Long Island, New York, shifting operations to Pompano Beach, Florida, in 1954. Fort Lauderdale's Striker was among a line of diesel-powered, aluminum-welded sport-fishing boats begun in 1961, and ranging long from 35 feet to 70 feet. During the Striker's twenty years of service, the sense of security it gave boat owners was offset by the damaging wake it created at marinas. In 1999, Fort Lauderdale Fire-Rescue placed in service a 25-foot Boston Whaler as Fireboat 49. Rated at 750 gpm, the vessel served until April 2004, when its pump died fighting an intense blaze aboard a freighter. Replacement Fireboat 49 was a 28-foot McKee Craft 1,000-gpm fire-rescue boat placed in service in November 2004. Located at Fairmont, North Carolina, Mckee Craft bills its boats as unsinkable, thanks to a process that pressure-fills the entire deck and hull with foam, forming an unsinkable, one-piece boat, eliminating the need for spars and stringers. The unique process was discovered by Lanness McKee in 1996. Fort Lauderdale Fire-Rescue's McKee Craft is a Model 28CC (Center Console) Freedom Fire Rescue. Twin outboard engines give Fireboat 49 a top speed of 55 mph.

The first fireboat known to have served the fire and police departments of Fort Lauderdale, Florida, was this 28-foot SaftiCraft nicknamed Old Reliable. It could pump 1,500 gpm, with delivery accomplished through Mystery deck monitors and five discharge outlets. Author's Collection.

After purchasing this 37-foot Striker sport fishing boat in 1970, the Fort Lauderdale Fire Department converted it to a fireboat, which served for twenty years. Courtesy of Fort Lauderdale Fire & Safety Museum.

Fort Lauderdale's 28-foot McKee Craft is of one-piece, solid foam construction. The center-console model is rated at 1,000 gpm and has a top speed of 55 mph. Author's Collection.

The bow monitor of Fort Lauderdale's Striker fireboat is brought to bear on a waterfront warehouse blaze. Courtesy of Fort Lauderdale Fire & Safety Museum.

Fort Lauderdale firefighters advance a hose line aboard burning Canadian vessel *Newfoundland Explorer* on April 25, 2004. Hose lines initially taken below deck caused a violent steam explosion; however, ten firefighters escaped. Here, they renew their attack on the listing yacht with a hose from Fireboat 49 up close in a sea of foam used to fight the blaze. Courtesy of Fort Lauderdale Fire & Safety Museum.

"Think Amphibious"

It was the catchphrase echoed by designers and builders of a unique amphibian that combined the features of fireboat and fire truck. The Miami Fire Department claimed the first commercial version of this highly effective and cost-saving concept that had the potential to revolutionize the thinking of fireboat design. Miami features the largest passenger seaport in the nation, plus more than twenty miles of shoreline on Biscayne Bay, the Miami River, and numerous offshore islands. The Miami Fire Department and the Coast Guard spent a great deal of time transporting firefighters to these islands to battle brush and forest fires, which often resulted from recreational use of the islands. With the ever-present threat of hurricanes, rugged and versatile amphibians are an obvious choice for fire and rescue. In the wake of Hurricanes Katrina in 2005 and Sandy in 2012, officials lamented the lack of amphibious fire-rescue vehicles. All considered, including Miami's growth and the busyness of its port, this teeming coastal metropolis was long overdue a fireboat; during the late 1970s, however, change was forthcoming.

In 1978, Louis D. DeChime was appointed division chief of support services and superintendent of the fire department's garage. Since the Miami Department of Fire, Rescue, and Inspection Services had been searching for an unconventional solution to the problem of meeting its fireboat needs, DeChime's first assignment was to identify a design for Miami's first fireboat. It was critical not only that the cost not exceed a bond program implemented twenty years earlier, time was of the essence since the bond was due to expire. The bond became a moot point after DeChime's fact-finding tour of several cities with large fireboats revealed that the budget was severely underfunded. Funds for a fireboat that met the city's needs included the cost of dock space and staffing a large boat. Deemed unfeasible, the fireboat plan was dropped. But DeChime remained diligent, and when he became aware of NASA's high-tech Firefly pump, the wheels began to turn. And when he was invited to ride Coral Gables' military surplus LARC amphibian, the wheels turned faster. He was impressed with the vehicle's ability to remain stable under a 5-ton load while negotiating rough terrain, including a 60-degree incline, and handle 12-foot waves. DeChime clearly visualized combining the LARC and the Firefly as Miami's first fireboat. It presented as a no-brainer when it became quickly apparent that an amphibious fireboat was within budget, could be cross-manned with existing crews, and could be housed on land, facilitating ease of maintenance. In addition, the unit could respond to land and water incidents and save valuable time traveling overland to a launch point nearest a water emergency. This was especially important on the Miami River where a no-wake law ruled out fast response. DeChime presented the plan to Chief Herman W. Brice, who gave the green light for the project.

DeChime went immediately to work, first gaining approval for the conversion to be done at the fire department garage, and then working with NASA officials to arrange transfer of a LARC from Fort Story, Virginia. The next step was purchasing from Aviall in California one Firefly unit to evaluate the total system to conduct a feasibility demonstration in the mouth of Biscayne Bay. The LARC arrived in 1979 and was loaded with ballast to duplicate the fireboat fully equipped. Through all phases of the evaluation in rough seas, the vessel never exceeded a 6-degree roll, and the Coast Guard rescue boat standing by was not needed. During the demo, Chief Brice asked DeChime, "You sure this is going to work?" When DeChime assured him it would, Brice added, "If it doesn't, we're both through here!" The demo far exceeded expectations, Brice and DeChime kept their jobs, and NASA agreed to arrange transfer of two additional LARCs from the army for a second fireboat and one for spare parts.

Under the direction of DeChime and assistant Winskie, conversion of the LARC to fireboat began during late 1980, with manufacturers of components on hand to install their wares. The Firefly II pump was chosen, which would deliver 3,000 gpm at 150 psi or 4,200 gpm at 100 psi. Miami shop personnel designed unique water intake boxes mounted to both sides of the hull, along with distinctive plumbing that fed a specially built Santa Rosa 2,900-gpm monitor on the rear deck; 360-degree rotation of two 1,100 gpm bow monitors allowed them to be used as thrusters to aid steering when operating the massive rear monitor. One dozen discharge outlets were included to supply land-based units. The cabin was enclosed, sound-proofed, and air-conditioned.

When completed, the amphibious fireboat, named Fireboat 3, became operational in December 1982 and was assigned to Miami's Engine Co. 3 and Foam 3 of the department's Special Tactics Team. The vessel was well received by crewmen who learned to steer the unit to compensate back pressure of the powerful monitors. The massive 6-inch monitor alone easily propelled Fireboat 3 at 8 mph, and its knockdown punch at fires was equally impressive.

So firm was DeChime's belief in the future of the amphibious fireboat that in January 1983 he retired from the Miami Fire Department to pursue the venture. When he convinced the president of LTI to produce the commercial version of Miami's Fireboat 3, the firm took him aboard to assist with the project. He then went into partnership with Robert Milnes of newly formed Firefighting Innovations to market the amphibious fireboat. All was blue skies and fair winds, with the first LTI fireboat delivered to Milwaukee in mid-1984. At the same time, the Miami Fire Department honored NASA's request to demonstrate Fireboat 3's capabilities at its Marshall Space Flight Center. Tampa Fire Department took delivery of the second unit in 1986. The first clouds over the future of the amphibious fireboat gathered early in 1986 when the British firm Simon purchased LTI, halting immediately the manufacture and sales not only of fireboats, but of pumper trucks, to concentrate solely on aerial apparatus. Fortunately, new possibilities for the amphibious fireboat loomed in New Jersey and New York.

Miami's Fireboat 3 served into the 1990s, when it went to Miami's sister city fire department of Cali, Colombia. It wasn't until 2007 that the Miami Fire-Rescue Department followed the lead of neighboring Miami-Dade County—which placed in service a 50-foot MetalCraft Marine fireboat—by ordering a MetalCraft Marine Firestorm 48, which began service at Miami Port in April 2009. Drafting only 2 feet and able to reach a speed of 44 mph , the all-aluminum Fireboat 1 is powered by twin 760 hp diesel engines driving Hamilton jets and delivering 4,000 gpm through twin Hale pumps; a 5-inch Storz discharge outlet was provided to supply two 1,750-gpm land pumpers. The cabin includes provisions enabling patient care and long periods as a command center.

The Miami-Dade Fire Rescue Department comprises sixty-five stations providing fire protection and EMS to unincorporated sectors of Miami-Dade County, along with thirty municipalities within the county. The department's Marine Operations' primary vessel is Fireboat 1, the 50-foot Firestorm, which is powered by twin Caterpillar 885 hp diesel engines driving Hamilton waterjets and twin Hale 3,500-gpm pumps. Fireboat 2 is a Firestorm 36 powered by twin 540 hp Cummins diesels driving waterjets and twin Darley 1,750-gpm pumps. The boat also features a 70-gallon foam tank. Other vessels in the fleet include two 28-foot Rigid Hull Inflatable Boats (RHIB), a 24-foot RHIB, seven various size RHIBs, an airboat, and ten Jon boats.

Miami's amphibious Fireboat 3 began service in 1982. Among its many features was a polished aluminum hull and specially built main monitor rated at more than 3,000 gpm. The monitor was positioned over the propeller and rudder for optimum control of its tremendous back pressure. All three monitors were remote-controlled and electro-hydraulically operated. Courtesy of Bob Milnes.

Fireboat 1 of Miami Fire-Rescue was built with a spacious non-skid afterdeck conducive to dive operations. Courtesy of MetalCraft Marine.

Miami's Fireboat 1 sails in choppy seas in Miami Port shortly after beginning service in 2009. The MetalCraft Marine Firestorm 48 has a top speed of 44 mph. Courtesy of MetalCraft Marine, Inc.

Miami-Dade's Firestorm 50 in its original scheme. Miami-Dade Fire Rescue apparatus began appearing in lime green livery in 1974. Bow and roof monitors were each capable of delivering 2,000 gpm. Courtesy of MetalCraft Marine, Inc.

Miami-Dade's Fireboat 1 in its operational scheme. Standard on MetalCraft Marine fireboats are push knees edged with rubber fendering, a stern dive platform, and raked windshield, which reduces glare and improves water run-off. Courtesy of MetalCraft Marine, Inc.

Miami-Dade's Fireboat 1 keeps pace with Air Rescue 2, one of four Bell Model 412 helicopters operated by the department's Air Rescue Bureau. Being multi-mission, Miami-Dade's marine and air units commonly work hand-in-hand. Courtesy of MetalCraft Marine, Inc.

Miami Fire Department's amphibious fireboat was featured in this advertisement during the 1980s. Courtesy of Bob Milnes.

Both Firestorms arrived in 2007; however, by 2012 they were removed from service in the wake of political turmoil between the union and city officials. In October 2014, both were placed back in service following a Fourth of July boating accident that left four dead. Their return to the water marked the first step in establishing a joint maritime response program for South Florida waters that included Miami-Dade Fire Rescue's two boats, Miami's Fireboat 1, and Miami Beach's vessels, with Coral Gables and Key Biscayne joining later. After only three months, Miami-Dade leadership pulled the plug on the arrangement, citing firefighter staffing problems.

The helm and control center of Port Canaveral's Fireboat 1, a Firestorm 36. Courtesy of Ben Dottridge, Cotuit, MA.

Port Canaveral, Florida, fire rescue operates this 2011 Firestorm 36, which can deliver nearly 3,000 gpm. Courtesy of Ben Dottridge, Cotuit, MA.

CHAPTER 5: GULF OF MEXICO

Throughout history, the Gulf of Mexico has harbored the potential for emergencies that necessitate a significant marine fire-rescue presence. The US states that border the Gulf—Florida, Alabama, Mississippi, Louisiana, and Texas—feature more than 1,600 miles of coastline, not including bays and inland waters. Maritime traffic abounds not only at some of the world's busiest ports along America's Third Coast, but in conjunction with flourishing resources of commercial and recreational fishing, and natural gas and oil sites.

Tampa

The sprawling, heavily populated metropolis of Tampa and the Tampa Bay Area dominates Florida's west coast. The city of Tampa established its paid fire department on May 10, 1895, and purchased its first motorized apparatus in 1909. Although Tampa's port lay in a large bay area, a decade would pass before city officials considered acquiring a fireboat. An additional four decades would pass before a fireboat actually appeared in Tampa Bay. That occurred in 1961 when Tampa placed in service the 65-foot *Gail Holland,* namesake of the daughter of Senator Spessard Holland, who assisted in obtaining the vessel.

Built by Calumet Shipbuilding & Drydock Company of Chicago, the *Holland* was one of eight steel-hull, diesel-powered Distribution Box Boats the yard turned out during 1943 for the US Army's mine-laying fleet. The Tampa Fire Department installed a 2,000-gpm pump to supply monitors on platforms at the bow and atop the pilothouse.

Similar to and replacing the *Gail Holland* was the 68-foot *Francis Bellamy,* built in 1956 by Paasch Marine of Erie, Pennsylvania. It was christened as Tampa's fireboat on April 16, 1974, in honor of former Tampa

The Port of Tampa's fireboat *Gail Holland* began life as a Distribution Box Boat for the army's mine laying fleet during WWII. The 65-foot, 2,000-gpm vessel mounted a boom on its afterdeck for a small launch. The fixed monitor tower atop the pilothouse was heightened from an earlier version. Courtesy of Akira Suwa.

Tampa's fireboat *Francis Bellamy* was similar in configuration to its predecessor, the *Holland.* The boat was painted overall white with a blue band on its funnel. Curt Elie Collection.

Following Milwaukee's example, Tampa Fire Department officials worked closely with LTI designers to convert a military surplus LARC V into an amphibious fireboat. Rated at 2,500 gpm, the amphibian began service in 1986. Courtesy of Bob Milnes.

resident Francis J. Bellamy, who penned the Pledge of Allegiance. Besides the two monitor platforms salvaged from the *Holland,* the *Bellamy* was fitted with two additional monitors to take full advantage of the boat's 6,000-gpm rating. The *Bellamy's* arrival signaled a shift in the location of smaller rescue boats from Tampa's Garrison Channel to Station 17 on Davis Islands.

Having monitored closely the results of Milwaukee's LARC fireboat conversion by LTI, Tampa fire officials worked with the firm to produce a nearly identical amphibious fireboat to augment Tampa's waterborne firefighting capability. Designated Fireboat 2 and nicknamed *Duck*, the 2,500-gpm amphibian arrived during late 1986 to provide Tampa's Bravest with a versatile, rapid land and water firefighting capability. Soon after beginning service, Fireboat 2 proved its worth by driving city streets to a point close to a shrimp boat burning off the coast, and then launching to sea. The amphibian reached the shrimper in sixteen minutes, versus the two hours officials estimated it would have taken a conventional fireboat navigating deep water.

Nearing its twilight, the *Francis Bellamy* was replaced in 1991 by Fireboat 1, a 30-foot, triple-engine-powered SeaArk rated at 1,250 gpm, and capable of reaching 46 mph. Later the SeaArk was renamed *Richard D. Santmyer* in honor of former Tampa Fire

Monitors on the foredeck of the amphibious fireboat were remotely controlled. Tampa's Fireboat 2 was appropriately nicknamed The Duck. Courtesy of Tampa Fire Rescue.

A 32-foot SeaArk Marine fireboat makes a test run in 1991 before delivery to the Tampa Fire Department. The boat is rated at 1,250 gpm. Courtesy of SeaArk Marine.

After delivery and christened *Richard D. Santmyer*, the boat's twin Evinrude engines were replaced with triple Mercury outboards, its bow monitor was replaced by an improved model, and it was repainted. Courtesy of Tampa Fire Rescue.

Department Special Operations Chief. A US Coast Guard reservist and supervisor of the Marine Division, Santmyer worked with Tampa's Captain of the Port to develop the *Burning Ship Contingency Plan* and coordinated large-scale exercises. The *Santmyer* was berthed at Davis Island's Marjorie Park Marina.

When the *Francis Bellamy* was sold to Florida's International Ship Company, its monitors and conspicuous tower were removed and installed on the tug *Gulfcoast,* owned by Gulf Coast Transit, a subsidiary of Tampa Electric Company, known as TECO. Tasked with maneuvering combination tug-coal barge operations at TECO power plants in Tampa Bay, *Gulfcoast* also served as a fireboat, assisting Tampa Fire Department's Marine Division formed in 1992. In fact, the day *Gulfcoast* left the shipyard where it was outfitted for firefighting, it worked at one of the largest maritime fires in Florida's history.

On August 10, 1993, the blaze engulfed a phosphate-loaded ship and two massive barges carrying petroleum products near Fort Desoto Park at the mouth of Tampa Bay. Working with the amphibious fireboat, *Gulfcoast* pumped for more than thirty hours, applying both water and foam on the blaze. The vessel's military surplus pump, rated at about 2,000 gpm, had been mated with a Caterpillar 3208 diesel engine.

The *Gulfcoast* was stationed twenty miles south of downtown Tampa at TECO's Apollo Beach facility, its crew having undergone an extensive firefighting training course. On board were 2,000 gallons of foam and 250 gallons of AFFF. With a crew of only two, the *Gulfcoast's* sea chest and wheelhouse monitor was controlled from the wheelhouse.

To bring its marine firefighting capability up to twenty-first century standards, Tampa Fire Rescue in 2009 purchased a model Firestorm 70 from MetalCraft Marine. Considered to be the fastest and the most powerful fireboat in the nation for its size, the 52-ton vessel, named *Patriot,* gave Tampa not only a high-speed, high-capacity pumping station rated at 12,000 gpm, its extended deck house serves as a command center and patient care unit.

Tampa Fire Rescue's Fireboat 1, the *Patriot*, is a study in power afloat, having a rating of 12,000 gpm, making it the most powerful fireboat for its size. The pilothouse monitor alone can deliver 5,000 gpm. Courtesy of Ron Frehm.

At the heart of the *Patriot* are Quad Iveco 825 hp diesels driving Quad Hamilton 364 waterjets and two Hale pumps to achieve high-volume water output and speeds exceeding 45 mph. *Patriot's* remotely controlled Stang monitor atop the wheelhouse alone delivers a 5,000-gpm knockout punch, while two bow monitors each deliver 2,000 gpm.

Special features include a crane on the deck house roof for deploying a small boat. On *Patriot's* bow is MetalCraft Marine's signature push knee, which not only incorporates a ladder and rubber fendering, but acts as a stabilizing float in high seas. Tampa Fire Rescue covers the port area, shipping channel, and all waters of Tampa Bay south to Egmont Key. To accomplish this, the Marine Division and Port Authority operate not only the *Patriot* and *Santmyer,* but a 27-foot, 1,250-gpm Boston Whaler and several shallow draft boats. The Marine Unit of nearby St. Petersburg Fire and Rescue operates a 780-gpm Boston Whaler Vigilant providing additional protection in the region.

The *Patriot* measures just less than 70 feet, with a beam of 23 feet, and drafting less than 3 feet. The wide stern recovery deck is low to the water to facilitate dive operations. The *Patriot* is manned by Station 17 on Davis Island. Courtesy of Tampa Fire Rescue.

The firefighting capability of Tampa Electric Company's tug *Gulfcoast* was enhanced with the monitors and conspicuous tower salvaged from the fireboat *Bellamy.* The molded rubber bow fender and "H" bitts mark its role as a tug. Crumpled pilothouse visors often were the result of close-in contact with the upward flare of a ship's hull. Overall white replaced its buff scheme in 1990. Courtesy of TECO Energy.

St. Petersburg, Florida, fire chief Claud Nesbit looks over his department's new fireboat, which was turned over by the local Coast Guard auxiliary in July 1945. The 30-foot, 800-gpm Hanley boat soon was nicknamed Suicide Susan. Courtesy of US Coast Guard.

Mobile, Alabama

Northward of Tampa along the Gulf coast, a number of municipalities have become equipped with fireboats. West of Florida's panhandle, Mobile, Alabama, along western Mobile Bay has long enjoyed fireboat protection. But despite growth of the deepwater port region since WWII, the city did not acquire a fireboat until two decades after the war.

In 1965, the Mobile Fire Department acquired the tug *Captain Wilson,* which had been operated by Mobile Towing & Wrecking Co. Already well equipped with fire pump and four monitors, the tug was repainted and renamed *Lurleen* for Alabama Governor Lurleen Wallace. During late 1966, to provide the city of Mobile economic relief, the fireboat was placed into the hands of Alabama State Docks, which kept a pilot and engineer aboard for around-the-clock response.

In 1974, General Services Administration donated to the Mobile Fire Department the 64-foot army mine planter boat (L-80), which actually was a Distribution Box Boat, or "L" boat, which serviced electrical junction boxes of mine fields. It was one of ten steel "L" boats built by Luders Shipbuilding, Stamford, Connecticut, during 1940 and 1941 under Design 209. The Mobile Fire Department outfitted it to pump 3,000 gpm and named it the *Ramona Doyle* in honor of the wife of former Mobile Commissioner Robert Doyle. The *Doyle* is best remembered for the life-saving actions of its crew on September 22, 1993, when a passenger train derailed and plummeted into the Mobile River Delta Bayou, killing forty-seven and injuring more than 100 people.

As it plied the waters of Mobile Bay, Alabama, the tug *Captain Wilson*, of Mobile Towing & Wrecking Co., seen here in 1958, was easily mistaken for a fireboat. The tug was so well equipped for firefighting that in 1965 it was purchased by the Mobile Fire Department to fill the role of fireboat. Author's Collection.

The *Doyle* was decommissioned in 2003, and in 2006 the Mobile Fire Rescue Department placed in service the first fireboat built to its specifications. Named *Phoenix,* the 40-foot fireboat pumps 2,000 gpm.

Fireboats that help protect Mississippi's Gulf shoreline are operated by the Biloxi and Gulfport Fire Departments. Thanks to a grant from BP Public Limited Company, Gulfport in 2010 took delivery of a 30-foot, 3,000-gpm "Argus" built by Seattle's Northwind Marine. When not escorting liquefied natural gas tankers, the tugs *M/V Signet Stars & Stripes* and *M/V Signet Constellation* of Angola LNG Supply Services are home

Pumping at full capacity through its four monitors, Mobile's fireboat *Lurleen* closes with the burning Liberian refrigerator ship *Arctic Reefer* on August 26, 1965. Before the ship settled to the bottom of Mobile Bay, firefighters feared an explosion of the ship's fuel and refrigeration gasses. Courtesy of Curt Elie Collection.

ported at the Port of Pascagoula, Mississippi. Built in 2011 by Gulfport's Trinity Offshore, the high-tech tugs are equipped with Sea-Fire fire suppression systems, using two powerful monitors.

The *Romona Doyle* was given to the Mobile, Alabama, fire department after having served as an army "L" boat during World War II. It was decommissioned after twenty-nine years of service. As a fireboat, it is best remembered for its life-saving role following a passenger train derailment in 1993. Courtesy of Curt Elie.

The tugs *M/V Signet Constellation* (left) and *M/V Signet Stars & Stripes* of Angola LNG Supply Services are home-ported at Pascagoula, Mississippi. When not escorting LNG tankers, the high-tech pair help protect Mississippi's Gulf shoreline with their powerful fire pumps. Author's Collection.

New Orleans

Since the mid-1800s, the Port of New Orleans has reigned as the most significant Gulf coast port in the nation. "Crescent City's" port not only held the title, but by the 1940s, surpassed the amount of tonnage that passed through New York Harbor. Despite the steady climb in shipping, during the nineteenth century, the first commitment to waterborne fire protection became reality in 1875 when the New Orleans Harbor Protection Company took delivery of the fireboat *Protector*. The 158-gross-ton vessel was built as hull number 154 by Harlan & Hollingsworth of Wilmington, Delaware, one of the earliest iron shipbuilders. Furthering the commitment to port fire protection was the Port of New Orleans Dock Board, which in 1905 placed in service the fireboat *Samson*, built as a tug in 1881. Although mounting only one monitor, the 108-foot, iron-hulled *Samson* was rated at 6,000 gpm. With the *Samson* unable to keep pace with the marked increase in harbor tonnage, the Dock Board authorized construction of a new, powerful fireboat. Designed by New Orleans naval architect H. E. Cornell and built by Johnson Iron Works, Drydock, and Shipbuilding Company of New Orleans, the vessel was launched on May 12, 1923, and christened *Deluge* by Lynne Hecht, daughter of New Orleans Dock Board President. The May 13, 1923 edition of *New Orleans Times-Picayune* stated: "The Deluge . . . will be the largest and most powerful fire tug in the country. It will throw a stream of water 350 feet and its pumps can handle 10,000 gallons of water a minute, its capacity equaling that of twelve modern fire engines."

The *Samson*, built as a tug in 1881, was the second fireboat to serve the Port of New Orleans. The iron-hulled vessel stretched 108 feet, and its pumps could deliver 6,000 gpm. Painted overall red, the *Samson* lies alongside the twin-stack *Deluge* in 1958. Although a powerhouse in its day, *Samson* was unable to keep pace with the growth of New Orleans's waterfront and was removed from frontline service in 1964. Michael Calamari Collection.

The mighty fireboat *Deluge* in 1928, five years after it began service with the New Orleans Dock Board. The twin funnels corresponded to twin boilers of a single 1,200-horsepower steam powerplant. With the aid of electric generators, the engine drove four pumps, giving the *Deluge* a punch of more than 10,000 gallons per minute. Courtesy of National Park Service/Curt Elie Collection.

And until New York's pair of 140-foot jumbo fireboats began service in the twenty-first century, the *Deluge* was indeed the largest fireboat in America, stretching 139 feet long, with a beam of 29 feet, and drafting 14 feet, 6 inches. Displacing 370 tons, the all-steel *Deluge* featured a double-riveted hull. A 1,200-horsepower steam engine with twin boilers was fed by a 75-ton oil supply held in four tanks. Aided by electric generators, the engine allowed a top speed of 15 mph and drove four 450 hp multiple-stage centrifugal pumps, each rated at 2,600 gpm, for a total of 10,400 gpm at 225 psi. Pumps supplied four 7-inch monitors, one of which was mounted atop an aft tower, along with fourteen large-diameter discharge gates.

During a mutual aid training exercise in February 1955, the fireboat *Deluge* supplies water to three pumpers, including New Orleans Fire Department Rescue 1 and a Civil Defense engine. This view makes obvious the need to position fireboat monitors as high as possible. Courtesy of Louisiana Division/City Archives, New Orleans Public Library.

In 1962 the *Deluge* was re-powered with diesel-electric engines, which resulted in a single, shorter funnel. The "firefighting tug's" fixed tower, which positioned the monitor 35 feet above deck level, was retained throughout the vessel's long career. Awnings covering deck areas were common on vessels operating in hot climates. Author's Collection.

The *Deluge* served continuously for decades, sometimes averaging forty fires per year, which its crew fought bravely and skillfully. Throughout those years under ownership of the Dock Board, *Deluge* served in multiple roles, towing vessels and using its powerful propeller and monitors to wash silt from beneath docks and to flush riverbanks. Nearing forty years of service, the Dock Board deemed the *Deluge* too costly to operate under steam power. Since the boat was in good condition, converting it to diesel electric power proved financially sound, and in 1962 the job went to New Orleans' Avondale Shipyards. The *Deluge* was back on the job in November 1962 equipped with a 1,950 hp diesel drive engine and twin 1,700 hp pumping engines.

This powerful WWII-era tug served many masters before it went to the Puerto Rico Ports Authority in 1994. It served the army and navy before joining the *Deluge* in protecting the Port of New Orleans. Named *Lewis Bourgeois II*, its arrival in 1964 spelled the end of the 83-year old *Samson*. The tug's original crane was modified to function as a monitor platform. Author's Collection.

Two years later, the *Deluge* got an assist from a second fireboat, ushering out the eighty-three-year old *Samson*. The new addition to the port had been built in 1942 as hull number 504 by Ira S. Bushey & Sons of Brooklyn, New York, as a harbor tug. It went to the US Army as Port Crane LT-290, and just months later, was passed to the US Navy and re-designated YT-344 Swatane. In 1964 the vessel was sold to the Port of New Orleans and named *Lewis Bourgeois II*. Unique among the vessel's firefighting equipment was a piped waterway and monitor incorporated into an articulating boom as a modification of the boat's original crane. The 94-foot *Bourgeois* was powered by an 805 hp diesel engine.

After the Dock Board sold its inspection yacht *Good Neighbor* in 1966, inspection and goodwill tours were added to the list of duties performed by the two fireboats. Together, the *Deluge* and the *Bourgeois* wrote many pages of firefighting history in New Orleans. In April 1969, their brave crews were credited with saving the port from disaster when an oil-laden barge cut across the bow of the tanker SS *Union Faith,* cutting the barge in half and setting both vessels and the river ablaze. Another hero was the tanker's captain, Crescent River Pilot Kenneth H. Scarborough. He was last seen racing through the flames toward the bow where he dropped both anchors, ending the flaming ship's drift toward the waterfront. Twenty five other crewmen were killed.

Nearly seventy years old, the *Deluge* in 1992 was placed on reserve status pending the arrival of a new fireboat ordered by the Port of New Orleans. In the meantime, the US Coast Guard donated to the New Orleans Fire Department a 30-foot UTM (Utility Boat, Medium), a fiberglass-reinforced plastic boat built during the 1960s by Pearson Corp. of Bristol, Rhode Island. Designed for sheltered waters, the boat's response area included more than 600 square miles of Lake Pontchartrain and three marinas. Named *Phoenix,* the boat was dedicated on December 21, 1993. The *Phoenix* was equipped with a 500-gpm pump and a roof monitor for foam or water. The new fireboat was named the *General Roy S. Kelly,* in honor of retired Brigadier General Roy S. Kelly, who became chairman of the Board of Commissioners of the Port of New Orleans and died in 1993. The 95-foot, aluminum boat was built as hull number 1057 by Bender Shipbuilding and Repair Company, Mobile,

The New Orleans Fire Department fireboat *Gen. Roy S. Kelly* doubled as a passenger vessel for port officials. A 3,000-pound-capacity crane on the aft deck gave the boat even more flexibility. Author's Collection.

The *Gen. Roy S. Kelly* of the New Orleans Port Authority demonstrates its firefighting capability in 1996. Among six monitors are two bow monitors for fires under the Crescent City's typically high waterfront structures. Courtesy of Curt Elie.

Alabama. The *Kelly* arrived in 1994 and began operations with a crew of four who had received fire and rescue training.

Two of the vessel's four 900 hp diesel engines drive the boat's conventional propellers, while two power waterjets assist drive and power bow thrusters. The boat's 13,500-gpm muscle power is distributed to four 2,500-gpm Stang monitors and two 1,000-gpm under-wharf nozzles that work in conjunction with two on the transom. Twenty-one discharge gates are installed, three of which can be used to deliver 2,000 gallons of AFFF concentrate. A glance at the *Kelly* makes obvious its multi-purpose function, like its predecessors, which includes serving as tour boat for shipping executives. Tours of port facilities often are catered affairs in the main deck air-conditioned cabin, which accommodates twenty-four passengers.

While the historically significant *Deluge* slipped into obscurity, eventually passing into private ownership and becoming derelict, the New Orleans Fire Department in 2011 placed an order with Moose Boats for a 36-foot, 1,500-gpm fire-rescue boat powered by twin 350 hp outboards. Named *Blaze,* the high-speed boat is a compact mono-hull platform of a new line of Moose boats, which is a deviation from the firm's signature catamaran hull.

The Port of South Louisiana stretches fifty-four miles along the Mississippi River between New Orleans and Baton Rouge, and is headquartered at Laplace, Louisiana. As one of the largest bulk cargo ports in the world, the port commission relies on an Emergency Response Division, which includes fireboats. The port's first fireboat was the *John James Charles*, named

The Port of South Louisiana's first fireboat was this converted sea-going trawler named the *John James Charles*, after the three parishes comprising the port. Author's Collection.

Under the parent Lower Mississippi Port-Wide Security Council, the Port of South Louisiana has a full-time emergency response division, which joins forces with four lower Mississippi River deep water ports to cover 225 miles of the river. Included are these Port Support Vessels, the 57-foot *Responder* (foreground) and the 48-foot *Accardo*, which were built by SeaArk Marine. Courtesy of SeaArk Marine, Inc.

Three fireboats are among the protective services provided by the Plaquemines Port, Harbor, and Terminal District. The *Authority I* and *Authority II* are twin vessels rated at 1,500 gpm. Training with other protective agencies is conducted regularly. Here, *Authority I* trains with a HH-65C Dolphin helicopter based at Coast Guard Air Station New Orleans. Courtesy of John L. Pennison/Plaquemines Port Harbor and Terminal District.

The *Authority III* of the Plaquemines Port, Harbor, and Terminal District is a powerhouse capable of delivering 15,000 gpm. A substantial foam capability is also possible with onboard storage of 1,000 gallons of foam concentrate. Despite its size, only a crew of three is necessary to operate the vessel. The *Authority III* was built by Conrad Shipyard, Morgan City, Louisiana. Courtesy of John L. Pennison/Plaquemines Port Harbor and Terminal District.

The seismic research vessel *Jeff Chouest*, then owned by Edison Chouest Offshore (ECO) of Galliano, Louisiana, demonstrates its master stream capability as it cools a semi-submersible offshore rig during a flare test in December 2006. ECO vessels fought the intense fire in the critical first hours of the Deepwater Horizon rig on April 20, 2010. The 177-foot *Chouest* was built in 1981 by North American Shipbuilding of Larose, Louisiana. Photo by Mieko Mahi/courtesy of Diamond Offshore Company.

Seven Platform Supply Vessels (PSVs) demonstrate the tremendous power of their fire pumps at the burning and collapsing offshore rig Deepwater Horizon on April 21, 2010. Courtesy of US Coast Guard.

for the port region's facilities: St. John the Baptist, St. James, and St. Charles Parishes. Joining the *John James Charles* in 2006 was the 48-foot Port Security Vessel *Accardo,* built by SeaArk Marine. In March 2007 the commission added the *Responder,* also by SeaArk. In mid-October 2015, the Port of South Louisiana took delivery of one of the most advanced fireboats afloat, a 75-foot, multi-purpose catamaran powered by twin Cat C-18 diesel engines. Separate powerplants drive pumps that deliver 6,000 gpm through three radio frequency-controlled monitors, and through an onboard hydrant system. Under auspices of the Lower Mississippi Port-Wide Security Council, the Emergency Response Division joins forces with four lower Mississippi River deep water ports to cover 225 miles of the river.

Tasked with protecting the gateway to upriver ports and the entire Mississippi Valley Export Corridor, the Plaquemines Port, Harbor and Terminal District operates three substantial fireboats. The agency is headquartered at Braithwaite, Louisiana, and has jurisdiction over the first 100 miles of the river that flows through Plaquemines, protecting its residents and the nation's two largest coal terminals. Since 1983, two 50-foot fire, rescue, and patrol boats have served the District. Deputy Port Director John L. Pennison offered:

> THE *AUTHORITY I* AND *AUTHORITY II* HAVE DONE EVERYTHING FROM FIGHTING FIRES AND TOWING A DISABLED FERRY BOAT TO TRANSPORTING IMMIGRATION AND CUSTOMS AGENTS FIFTY MILES OFFSHORE TO MEET SHIPS. DURING HURRICANE KATRINA, THEY ANSWERED THE NEEDS OF HURRICANE VICTIMS.

The twin fireboats, built by Swiftships of Morgan City, Louisiana, are rated at 1,500 gpm and carry 300 gallons of foam concentrate. Each is operated by a crew of two.

Pennison continues:

> AFTER THE TERRORIST ATTACKS OF SEPTEMBER 11, 2001, THE PORT WAS GIVEN A $3 MILLION HOMELAND SECURITY GRANT TO DESIGN AND CONSTRUCT A NEW MULTI-PURPOSE VESSEL. THE PORT TOOK DELIVERY OF THE ALL-ALUMINUM 90-FOOT SECURITY VESSEL IN DECEMBER 2005. THE *AUTHORITY III* USES THREE CATERPILLAR C32 E-RATED ENGINES, WHICH DEVELOP 1,650 HP EACH THAT PROVIDE POWER THROUGH REDUCTION GEARS WITH A 2:1 RATIO. THERE ARE TWELVE FIXED SEATS FOR PASSENGERS OR CREW. MOUNTS AT EACH OF THE FOUR CORNERS CAN ACCOMMODATE AUTOMATIC WEAPONS SHOULD THE COAST GUARD NEED THAT CAPABILITY. MID-SHIP, THE MAIN DECK DROPS DOWN TO THE WATERLINE ON BOTH SIDES TO GIVE DIVERS EASY ACCESS TO THE WATER.
>
> THE MOST IMPRESSIVE CAPABILITY OF THE VESSEL IS FIREFIGHTING. THE MAIN FIRE PUMP CAN PUMP 10,000 GPM, AND A SECONDARY PUMP PROVIDES AN ADDITIONAL 5,000 GPM. A MANIFOLD SYSTEM PUMPS WATER ASHORE THROUGH HOSES; AFTER HURRICANE KATRINA, THE SYSTEM PROVIDED BACKUP COOLING WATER FOR CHEMICAL PLANTS AND REFINERIES. MOUNTED HIGH ABOVE THE WHEELHOUSE ARE TWO REMOTE-CONTROLLED MONITORS FOR DUAL-FLOW FOAM-WATER, ALONG WITH FIVE MANUALLY CONTROLLED MONITORS WITH FOAM INJECTION NOZZLES. ON *AUTHORITY III'S* BOW ARE REMOTE-CONTROLLED UNDER-WHARF MONITORS.
>
> AT THE 2006 INTERNATIONAL WORKBOAT SHOW IN NEW ORLEANS, THE *AUTHORITY III* WAS AWARDED AS ONE OF THE TEN MOST SIGNIFICANT BOATS IN THE NATION.

In October 2015, replacements for the District's *Authority I* and *Authority II* arrived from Metal Shark Boats of Jeancrette, Louisiana. Since the Gravois family, experienced custom boat builders, launched Metal Shark in 2005, the company has earned a reputation for welded aluminum boats custom built to customers' requirements. The new twins for Plaquemines Port are Metal Shark 55-foot Defiants, which are multi-purpose port security and fire-rescue models.

It took a major disaster forty-one miles off Louisiana's coast to spotlight the firefighting capability of platform supply vessels (PSVs). On April 20, 2010, an explosion and fire on the Deepwater Horizon oil-drilling rig sent flames 200 to 300 feet in the air. During two days before sinking in the Gulf, PSVs surrounded the doomed rig to pour thousands of tons of seawater on the inferno. Platform supply vessels, which range up to 350 feet long, commonly are outfitted with powerful firefighting pumps that supply monitors atop their towering superstructures.

Meanwhile, in Texas

Since much of the material handled in ports along the Gulf of Mexico consists of petroleum products, refineries and bulk storage and handling facilities often are self-protected against fire and explosion. Such was the case with the diesel tug *Delanco*. Weighing 183 gross tons, the 94-foot boat was built in 1920 by Johnson Iron Works of New Orleans. The *Delanco* served as fireboat for Texas's Port Arthur Refinery, the largest oil refinery in the nation, until 1983, when she was purchased by Gulf Oil Company.

During the months following the 1947 Texas City disaster, in which more than 500 lives were lost when the ammonium nitrate-laden ship SS *Grandcamp* exploded, several chemical and refining companies in the Gulf region formed the Refinery Terminal Fire Company (RTFC). Funded by member companies, the RTFC employs more than 100 firefighters staffing seven fire stations, forming the largest non-profit industrial firefighting organization in the US. Unquestionably, the most remarkable apparatus operated by the RTFC is found at the Port of Corpus Christi.

Port of Corpus Christi commissioner Yolanda Olivarez christens *Port of Corpus Christi* on 6 January 2000. Author's Collection.

Lying along sediment-laden Corpus Christi Bay, the city of Corpus Christi teems with a population of more than 235,000, thriving as a major petroleum and natural gas production center. Heavy industry abounds, along with a major fishing industry, plus tourism on Padre Island, the largest barrier island in the US. In view of the port's steady growth, and frustrated with being ill-equipped to handle vessel and waterfront fires, in late 1998 Port of Corpus Christi officials teamed with Houston naval architects Schuller & Allen to design a unique firefighting vessel, plans for which had begun five years earlier. They chose a barge-type configuration that formed a 110-foot by 30-foot work platform. The barge was designed as a pumping powerhouse with most of its machinery intended for firefighting. The vessel is positioned by

Weighing 247 gross tons, Corpus Christi's firefighting barge offers an improved level of flexibility and versatility. Since nearly 90 percent of the port's cargo is flammable liquid, the barge is a good example of how firefighting vessels are designed to meet specific requirements. The centerpiece of the vessel is the three-section Schwing boom, which can reach inside a ship to pump foam, water, or dry chemical agents onto a fire. Courtesy of Ken Rudine.

on-call tugboats, with propulsion assist by hydraulic, retractable thrusters powered by a 420 hp Detroit diesel. The heart of the system is a massive, three-section Schwing boom featuring 90 feet of tip reach and a 400-foot stream reach; the boom's horizontal reach is 76 feet.

A pair of 4,000 gpm-pumps is driven by twin 660 hp Detroit Diesel AP-92 engines. Pumping capacity can be doubled by positioning two trailer-mounted pumps on deck. In addition to the boom, it has four 2,000-gpm electrically-controlled monitors. For safety, all nozzles can be operated from the aft pilothouse, or by radio control without anyone being aboard; 15,000 gallons of foam concentrate is stored in tanks for direct delivery through monitors. The port's decision to choose the boom was based as much on past performance as on future expectations. Since several RTFC trucks mounted Schwing booms and apparatus was powered by Detroit diesels, commonality was desired for ease of operation and maintenance.

Built by Steiner Shipyard of Bayou La Batre, Alabama, the barge-like vessel was christened *Port of Corpus Christi* on January 6, 2000, by Port of Corpus Christi Commissioner Yolanda Olivarez, and placed in service the following month. To avoid the expense of housing an onboard crew, deputy director of operations for the Port Authority Tony Alejandro explained, "We thought it would be far more beneficial to contract with personnel from the land-based RTFC and focus our efforts on a vessel with solely performance-focused features." Accordingly, thirty RTFC firefighters were trained for its operation.

To augment the *Port of Corpus Christi,* in early 2005, Bay-Houston Towing Company purchased the 102-foot, 3,500-gpm tractor tug *C-Tractor 1,* which it renamed *William M.* The firm has been in Texas waters since the late 1800s, providing the Port of Corpus Christi its first tugboats when it opened in 1926. All of Bay-Houston's tugboats are substantially equipped for firefighting, with its Z-tugs rated at nearly 11,000 gpm for fires at LNG terminals.

Galveston

At the beginning of the twentieth century, commerce of the Southwest was handled through the port of Galveston, Texas, ranked second only to New York as a shipping center. In view of the growth of the port and the city as a commerce gateway, marine officials considered Galveston fortunate to have escaped conflagrations such as those suffered by so many other major port cities. Unwilling to press its luck, the city teamed with marine and wharf businesses to lease the 100-foot tug *Charles Clarke* for conversion to fireboat. Muscle for the *Clarke* consisted of three pumps delivering a total of 4,200 gpm, all of which could be delivered through a large monitor on the after-deckhouse. According to the *Galveston Daily News*: "This monitor can hurl a stream from a 1.75-inch nozzle at a pressure of 350 psi—sufficient to unroof any of the large sheds on the waterfront, to tear away by force into the heart of a conflagration." At the monitor's base were eleven outlets to simultaneously supply land lines. A smaller monitor was mounted on the forward deckhouse.

The *City of Galveston* was the Texas city's first dedicated fireboat, which served twenty-two years before it was deactivated. The use of gasoline engines for both propulsion and pumping ruled out the need for large, bulky exhaust funnels. Courtesy of the Rosenberg Library, Galveston, Texas.

As a combination tug-fireboat, the *Clarke* frequently was unavailable, prompting marine underwriters in 1927 to recommend that a dedicated fireboat be purchased. A new vessel would be part of plans that called for new fire alarm and water main systems, and new land apparatus along the waterfront. The all-steel, 85-foot fireboat was designed by New York naval architects Cox and Stevens and built by Gibbs Gas Engine of Jacksonville, Florida. A pair of 500 hp, eight-cylinder Winton gasoline pumping engines far out-powered its two 125 hp Winton American engines that drove twin screws. Rated at 7,500 gpm, the vessel could throw powerful streams through four monitors, one of which was atop a fixed tower, plus supply two banks of discharge manifolds.

Galveston native and Florida naval architect John G. Hannah christened the fireboat *City of Galveston* before it slid the ways at Gibbs Shipyard in 1929; Gibbs was sold to Aerojet-General in 1964, then to Fruehauf in 1967, and closed its doors in 1989. The *City of Galveston* was stationed at Pier 22, manned by a dozen crewmen in two six-man shifts.

After twenty years of dedicated service, which included many large fires, the vessel's age and operating cost, plus new fire-resistant wharves, combined to force its deactivation. In 1952 the fireboat was sold for $500 to a dredging company, left derelict at its moorings, and eventually sank. Not until 1963 did Galveston fire officials formulate plans to acquire small rescue boats to save lives before state or federal units could arrive. In 2015, Galveston fire chief Michael W. Wisko lamented, "We do not currently have a fireboat. We have discussed it for the twenty-seven years I have been here, but it has never happened."

Houston

Growth of Houston's port earned it the title of nation's busiest in terms of foreign tonnage and its being home to one of the world's largest petrochemical complexes. In 1924, ten years after the twenty-five-mile-long Houston Ship Channel was completed, fire devastated the steamship *Utah,* prompting Houston Fire Commissioner A. Anderson's call for the ability to attack blazes from the water. Port officials in 1925 approved the purchase of a fireboat, which led to a contract with Bethlehem Shipbuilding of Wilmington, Delaware, for construction of a 118-foot-long vessel with a beam of 27 feet and drafting 8 feet. Designed by Cox & Stevens, this was labeled the nation's first fireboat powered by diesel electric engines; two Winton diesels connected to Westinghouse generators produced 2,360 horsepower for 8,000 gpm at 150 psi through five monitors. The advantages of diesel

Fixed steel monitor towers were common fixtures on large fireboats during the 1930s and 1940s. At the midpoint of *City of Galveston's* tower was a searchlight and operator platform. Immediately abaft her deckhouse were two banks of discharge outlets. Courtesy of the Rosenberg Library, Galveston, Texas.

On November 8 and 9, 1959, an explosion and fire aboard the tanker *Amoco Virginia* threatened the nation's largest oil refining center. The vessel's captain and six crewmen died in the blaze at Hess Terminal. The fireboat *Capt. Crotty*, seen here nosed into the tanker, was assisted by a fireboat from the Houston Navigation District and three Coast Guard boats. A military airlift brought more than half a million pounds of foam to the site to bring the blaze under control. Courtesy of Curt Elie Collection.

The day that the *Capt. Crotty* began service marked the end of the Houston Fire Department's Marine Division, and the Port of Houston assuming control of fireboats. In 1976, the *Crotty* wore Bicentennial markings on the bow and pilothouse. Courtesy of Fire Apparatus Journal Collection.

The *J. S. Bracewell* was one of two identical high-speed, aluminum fireboats ordered in 1981 to improve fire protection for the Port of Houston. The multiple bows of monitors minimized turbulence of high-powered streams. Courtesy of Fire Apparatus Journal Collection.

electric over diesel were increased efficiency, greater reliability due to fewer moving parts, less space, and less weight. After Bethlehem launched its hull number 3500 at 284 gross tons in early 1926, the fireboat was named *Port Houston* and began service with the Houston Fire Department on 20 June.

When *Port Houston* reached the twenty-year mark, plans were formulated for a smaller, faster fireboat, which began life as hull number 200 at RTC Shipbuilding, Camden, New Jersey; the yard began simultaneous construction of an identical vessel for the government of Israel. The 74-foot vessel could pump 6,000 gpm and reach a speed of 20 mph. On March 20, 1950, Houston's new fireboat was christened *Captain Crotty* after assistant port director Charles Crotty. It was delivered on July 4, and when it began service on September 1, the Port of Houston assumed control of fireboats, effectively dissolving Houston Fire Department's Marine Division.

The *Crotty* responded to thirty-six alarms during its first year of service, most notably at the General American Tank Storage Terminal blaze in February 1951, where it pumped continuously for thirty-six hours.

During the 1950s, industries organized to pool their resources to respond to marine emergencies along the fifty-mile channel. Organizational changes during 1971 that gave the Port Commission added responsibility for firefighting paved the way for a new fireboat to augment the *Crotty*. To avoid the difficulty the *Crotty* encountered during its long voyage from New Jersey, the contract for the new vessel was let to Houston's own Bludworth Shipyards.

Since hazardous materials and oil spill responses were becoming the greater part of the Port of Houston Authority Fire Department's work, fireboats for the region had to be custom-designed as specialized apparatus. Accordingly, the new 80-foot boat was equipped with two 3,000-gpm pumps to deliver water, as well as other agents, through two 2,000-gpm monitors and an articulating boom with 1,000-gpm nozzle; the boom was later removed and replaced by a 1,500-gpm monitor. Also aboard were 1,800 pounds of dry chemical extinguishing agent and 2,000 gallons of AFFF, plus 3,000 feet of hose.

On April 15, 1974, the steel-hulled vessel was christened the *Capt. W. L. Farnsworth* by four-year old W. L. Farnsworth III, grandson of the boat's namesake. Captain Walter L. Farnsworth in 1914 navigated the first ocean-going vessel on the Houston Ship Channel. Driving the *Farnsworth* were two 1,250 hp V16 diesel engines, while 450 hp six-cylinder diesels powered its pumps.

At the beginning of the 1980s, manpower not only was increased, but marine firefighters began receiving more specialized training, and specialized land apparatus was added to the inventory. In its quest to maintain proficiency and continually improve, the Port Commission in 1981 decided to retire the *Captain Crotty* and accept bids for two high-speed aluminum fireboats. The identical 68-footers were built in 1983 by Stewart and Stevenson of Morgan City, Louisiana, and on May 26 were christened the *J. S. Bracewell* and the *Howard T. Tellepsen* after former port commissioners. Twin

Fireboat 2 of the Port of Houston Fire Department, the *J. S. Bracewell*, in 2010 with hull repainted high-visibility red, replacing its earlier black scheme. The *Bracewell* and its twin, the *Tellepsen*, mounted two manually operated 1,250-gpm monitors forward, while a 1,250-gpm monitor above the flying bridge was hydraulically operated. Dual hose reels on the afterdeck carried 1,600 feet of hose. Courtesy of Curt Elie.

The emblem of the Port of Houston Fire Department, Port of Houston Authority, which incorporates a twenty-first-century fireboat. Courtesy of Gordon J. Nord Jr.

The Port of Houston's Fireboat 2 was the first of three 70-foot Firestorms, having arrived in June 2013. The roof-mounted Stang monitor can throw a stream 450 feet. A 1-ton capacity crane atop the afterhouse hoists a Zodiac rescue vessel. The mast assembly incorporates radar, scene lighting, cameras, and search lights. Courtesy of MetalCraft Marine, Inc.

One of Port of Houston's three Firestorm 70s demonstrates its pumping power with far-reaching streams on the sunlit port. Courtesy of MetalCraft Marine.

Detroit diesel engines give the boats a top speed of 20 mph and power hydraulic motors geared to a pair of Gould fire pumps that deliver 4,000 gpm. Each boat carries 1,000 gallons of foam. The port now had three fireboats, each staffed by a crew of four, operating from stations at Manchester 3, Barbours Cut, and Bulk Materials Handling Plant on Greens Bayou. As land apparatus was added to the department early in the twenty-first century, Station 4 was opened at Bayport Container and Cruise Terminal.

A 2012 Homeland Security grant made it possible to modernize the fleet with three state-of-the-art Firestorm fireboats built by MetalCraft Marine. The Port of Houston Authority Fire Department took delivery of the first of the trio in June 2013, edging the *Tellepsen* into retirement. Christened Fireboat 2, the vessel was the department's first to utilize waterjet propulsion. More than 4,600 horsepower makes the 70-foot aluminum boat a study in speed and power; she can skim the surface at 45 mph and deliver an amazing 17,000 gallons of water per minute at 130 psi. As a shore hydrant, the boat's quad pumps allow a flow of 7,000 gpm at 70 psi through 1,000 feet of hose from a five-inch Storz outlet. Drafting less than 3 feet, the vessels can operate in the shallows of numerous islands along the channel. Enhancing the capabilities of the three fireboats are sophisticated electronic suites and the ability to operate in hazardous materials incidents and as a command center. After Fireboat 1 completed the two-week voyage through four great lakes and the Mississippi River in October 2013, the *Bracewell* was retired, with Fireboat 3's arrival in 2014 ushering out the *Farnsworth*.

The fireboat *Chief Randall J. Zimmer* began service with the Seabrook Volunteer Fire Department, Texas, in August 2012. The 3,000-gpm fire-rescue boat was designed by Northwind Marine for shallow water operations of Clear Lake and Galveston Bay. The *Zimmer* replaced the fireboat *Miss Lois*, a wooden 34-foot vessel that had served the region since 1985. Author's Collection.

CHAPTER 6: INLAND CITIES

St. Louis

Countless bodies of water and waterways across the nation are protected by fireboats. While none fall into the category of battleship-class vessels, some are significant in the history of large cities. The largest vessel known to have operated as a fireboat on inland waters hailed from St. Louis, which records the use of fireboats throughout much of its history. While not a pure fireboat, the massive, steel-hull side-wheeler was a combination fireboat, police patrol boat, and inspection-tour boat for St. Louis officials and visiting dignitaries. Measuring 186 feet long and 32 feet at its beam, the harbor boat's four boilers powered three engines capable of pumping 6,000 gpm through 2,000 feet of three-inch fire hose. This pumping power also allowed cleaning of the city's many levees.

The boat was designed by John Frederick Deppe, a machinist who was also skilled in boiler making and iron work. It was christened the *Erastus Wells* after a local businessman and politician, and father of St. Louis mayor Rolla Wells. Taken under tow at Grafton Boat Yards, Illinois, by the harbor boat *Mark Twain*, it arrived in St. Louis on October 16, 1907. The *Twain's* machinery was then removed and installed in the *Wells* and the *Mark Twain* was dismantled. The new harbor boat *Erastus Wells* was unique in that it could do 12 mph against the current of the mighty Mississippi, and was built with quarters to accommodate the master and half of the crew during emergencies.

Since city officials could not justify the boat's expense during the Great Depression, it was sold in 1932, converted to an excursion boat, and renamed *City of St. Louis*. For waterfront firefighting, the city, then the nation's second largest inland port, had to rely on vessels of the US Coast Guard, Army Corps of Engineers, and Bi-State Development Agency, plus tugs equipped with fire pumps. From the Coast Guard came a 30-foot Hanley Boat, which arrived by rail; the 70-foot *Percy F.* was acquired from the Bi-State Development Agency through federal surplus and converted for firefighting, and in 1967 the tug *Louise Leyhe* was outfitted for fighting fires. It was made available for fire service by owner Notre Dame Fleeting and Towing Service at no charge.

In August 1978 the Port Authority of the City of St. Louis purchased from the sewer district a 31-foot, steel-hull utility boat built in 1966 by Grafton Boat Works. It was named the *Capt. Targee* in honor of Captain Thomas B. Targee who lost his life fighting the Great Fire of 1849 on the St. Louis waterfront. Although outfitted with small 250-gpm gasoline-powered pumps and monitors from a 1928 water tower, the vessel was not up to the ardors of firefighting since sediment often clogged the pumps. Nor was its small inboard motor up to the task of maneuvering in the strong current of the Mississippi River. In fact, on the boat's first alarm to assist a vessel in trouble, its lack of power and inexperienced crew could not prevent the *Targee* from being swept under a barge. Luckily, the three crewmen were able to leap onto the barge, and a passing tug hooked onto the sinking boat and pulled it from under the barge. Although the boat was repaired, firefighters were reluctant to take it out.

The Port Authority of St. Louis owned the fireboat *Capt. Targee*, which was manned by St. Louis firefighters. The 31-foot, 1960s vintage steel boat was converted from a utility vessel during the late 1970s. It is seen here in 1980 temporarily equipped with NASA's Firefly pump. Courtesy of Bob Pauly, Curator, St. Louis Fire Department Museum.

In 1980, NASA loaned its Firefly pump module for the Port Authority of St. Louis and the St. Louis Fire Department to use in the Port Cities Demonstration Project. The Maritime Administration and NASA had purchased the first Firefly built, which delivered 3,000 gpm aboard the *Capt. Targee*. After an eight-month evaluation period, during which the vessel proved difficult to maneuver with the high-powered pump, the unit went back to NASA.

The small pumps were re-installed, and three years later the addition of a more powerful outboard motor only made the boat sluggish in the water. Lacking the ability to make foam, and being too slow for responses, the *Targee* fell into disuse and eventually was scrapped. Its replacement in 1992 was a 1977 24-foot MonArk 250-gpm boat donated by Union Electric. Manned by members of St. Louis Fire Department Engine Co. 11, the boat was named for the steamer *White Cloud,* whose onboard fire started the Great Fire of 1849.

In 1994 the Coast Guard donated to the St. Louis Fire Department a 32-foot Port and Waterways Boat that had been built in 1976 as part of a move assigning the USCG responsibility for inland waters. The all-fiberglass vessel was powered by two Caterpillar 3208 diesel engines and a Detroit 353 diesel fire pump delivering 500 gpm. Although designated Marine Unit 1, the boat was christened *Phelim O'Toole* in honor of an early hero of the paid fire department who had rescued twelve people from a hotel fire in 1877; O'Toole was killed fighting a fire three years later.

Tests with the NASA-loaned Firefly pump aboard the *Targee* gave the boat a 3,000-gpm capability, but created maneuverability problems. The Firefly was also flown to a barge in the middle of the river by a Sikorsky S-58 helicopter, and it then supplied fire monitors. Courtesy of Bob Pauly, Curator, St. Louis Fire Department Museum.

The Coast Guard gave the *Phelim O'Toole* to the St. Louis Fire Department in 1994. Fifty of the 32-foot Port and Waterways boats had been built from 1972 to 1976 to comply with the Port and Waterways Safety Act of 1972. The act, which came on the heels of a tanker collision in San Francisco in 1971, assigned the USCG with added responsibility for the protection of ports, harbors, and waterways. Courtesy of Dennis J. Maag.

The *White Cloud* was donated to the St. Louis Fire Department by Union Electric. The 1977 Monark boat pumped 250 gpm and was powered by twin outboards. The boat and trailer were pulled by Marine Unit 2, a 1988 Dodge truck. Courtesy of Dennis J. Maag.

The St. Louis Fire Department, in keeping with its reputation for using specialized equipment to protect its waterways in the twenty-first century, established a Marine Unit Task Force comprising four vessels. Leading the group is the newest and largest, the *Stan Musial,* a MetalCraft Marine Firestorm 45 capable of delivering nearly 8,000 gpm through four monitors. Twin 600 hp Cummins diesel engines drive Hamilton jets and two Hale pumps. Her sophisticated electronics suite includes radar, side-scanning sonar and Forward Looking Infrared Radar (FLIR).

Next is a 27-foot Boston Whaler Dauntless, named *Jack Buck* for a renowned sports announcer. Powered by twin Mercury 225 hp engines, the *Jack Buck* is tasked as a rescue boat but can hold its own fighting flames with a 500-gpm pump. The *Backstopper* is a 27-foot aluminum hull-rigid hull inflatable boat built in 2012 by Safe Boat in its Defender series. Primarily a rescue craft, the *Backstopper* is powered by twin 225 hp Mercury outboards and features a Darley 500-gpm pump for its bow monitor. The *White Cloud* is tasked with coverage of the northern port region near the confluence of the Missouri and Mississippi Rivers. Completing the task force are land units that include Engine Companies 11 and 27, Rescue Squad 1, Dive Truck, and two command chiefs.

Cincinnati, Ohio's *B. S. Wydman* was considered the oldest fireboat to provide fire protection on inland waters. Named after Byron S. Wydman of Cincinnati, the unit comprised a land steam engine mounted on a canal barge to serve as the city's first fireboat. The barge was built in 1885 and mated with the steamer for the

Framed by St. Louis' famous landmark, the fireboat *Jack Buck* in 2009 cruises the riverfront with officials aboard to determine the vessel's replacement. Courtesy of Bob Pauly, Curator, St. Louis Fire Department Museum.

The *Backstopper*, seen here in 2012, is a 27-foot fire-rescue vessel from Safe Boats's line of Defender models. Twin outboard engines provide nearly 500 horsepower, and the boat can deliver water at 500 gpm. Courtesy of Dennis J. Maag.

Cincinnati's *B. S. Wydman* was believed to be the oldest fireboat to have plied inland waters. Like many fireboats of the nineteenth century, the *Wydman*'s power to fight blazes came from a steam pumper secured to the deck. The *Wydman*'s dapper crew poses aboard the fire barge, which stood guard in the Miami-Erie Canal during the 1888 Cincinnati Exposition. Courtesy of Fire Museum of Greater Cincinnati.

1888 Ohio Valley and Central States Centennial Exposition to protect exhibit structures on the banks of the Miami-Erie Canal. The canal gave way to railway use and was later drained and paved. Cincinnati's next recorded use of fireboats lists three military surplus amphibious DUKWs converted to fireboats by fire department shops at the end of WWII. Later years saw the *Dr. Giles Decourcy,* a 1977 General Marine 4000, a 1981 MonArk 25-footer rated at 1,000 gpm and named *David T. Sheehy*, a 1991 SeaArk rated at 1,000 gpm and named *Carl H. Lindner*, and a 1998 27-foot Boston Whaler Challenger rated at 750 gpm and named *Herman Helfrich.*

One of the most powerful fireboats to ply inland waters was berthed at the south side of Pittsburgh, Pennsylvania. Rated at 10,000 gpm, the 70-foot all-steel fireboat was designed in 1954 and built by Paasch Marine of Erie, Pennsylvania. Although the shipyard was only 130 miles from Iron City, the boat could not be moved by rail or roadways, requiring a two-man crew to navigate 2,500 miles through the Great Lakes, down the Mississippi, and up the Ohio Rivers to deliver the boat. In 1956 it was placed in service by the Pittsburgh Bureau of Fire to cover the Ohio, Allegheny, and Monongahela Rivers. On April 21, 1957, the vessel was christened *C. D. Scully* in honor of Cornelius D. Scully, who served as Pittsburgh mayor from 1936 to 1946. After seventeen years of service, with many docks and warehouses gone from the riverfront, the *Scully* was deemed an unnecessary expense and sold to the Tampa Fire Department in June 1973. After fire aboard a ship pointed out the need for a marine firefighting capability, two 30-foot SeaArks of the River Rescue Unit of Pittsburgh EMS were equipped with 500-gpm pumps; Pittsburgh EMS is independent of the Bureau of Fire.

The Pittsburgh Bureau of Fire operated one of the most powerful fireboats on inland waters, the 10,000-gpm *C. D. Scully*, seen here testing its monitors on the Allegheny River at Point State Park during the late 1950s. The *Scully* went to the Tampa Fire Department in 1973. Author's Collection.

This fireboat was named *Al Fuentes* in honor of the FDNY Marine Division officer who was instrumental in getting the boat donated to the Memphis Fire Department in June 2008. The bow monitor is a 1,100-gpm Elkhart, which also delivered foam drawn from a 100-gallon tank. The Memphis Fire Department also operates a 26-foot 1,250-gpm SeaArk fireboat. Courtesy of Larry Shapiro.

The Nashville Fire Department joined the trend in fireboats from MetalCraft Marine in 2004 with this 43-foot, 3,000-gpm Firestorm. Fireboat 9 could reach a speed of 42 mph. Courtesy of MetalCraft Marine.

To protect its riverfront interests along the Mississippi, Memphis, Tennessee, placed a fireboat in service during the summer of 1984. The MonArk Model 2609 featured a welded aluminum hull housing three Cummins 6BT diesel engines, two for propulsion up to 27 mph and one for powering a Hale 1,200-gpm pump. In June 2008, the Memphis Fire Department received from the Fire Department of New York a 24-foot fiberglass-hulled Steiger Craft, which the FDNY had acquired in 2005 for work in Manhattan. Docked at the yacht club on Mud Island, the boat was named *Al Fuentes* after the FDNY officer who was instrumental in its donation.

The Cumberland River is another major waterway of the southern US. The largest city on the Cumberland is Nashville, Tennessee, whose fire department includes a Marine Operations Section charged with protecting nearly sixty miles of the inland waterway. To accomplish this, marine firefighters routinely board Fireboat 9, a 43-foot 2004 MetalCraft Marine fireboat rated at 2,900 gpm and capable of speeds up to 42 mph. The Marine Operations Section also operates Fireboat 22, a rapid response boat acquired in 2012 to replace a twenty-year-old Stratus boat. Built by Lake Assault Boats in Superior, Wisconsin, the 28-foot boat is rated at 1,100 gpm.

Throughout history, across the nation, municipalities both large and small have employed vessels of all types and sizes for fire-rescue duty on lakes and rivers. The Dallas, Texas, fire department, for example, has used boats on its lakes beginning with aluminum fishing boats during the 1950s. Other boats over the years included a deck boat, inflatable boat, and cabin cruiser, all of which fell short of providing complete fire and rescue service. Finally, in 2006 the department took delivery of a SeaArk named *Marine 1* in honor of New York's *Marine 1* and those who died at the World Trade Center on 9/11.

In January 2011, the Charlotte, North Carolina, fire department took delivery of a 30-foot fire-rescue boat built by Viking Welding & Fabrication of Kensington, New Hampshire. The 3,500-gpm, all-aluminum Fireboat 38 provides emergency response capability for Lake Wylie, and is one of five fire-rescue boats stationed on inland waters in the Tar Heel State.

Also from Viking Welding is a 31-foot fireboat, giving the St. Paul, Minnesota Fire Department a fire-rescue capability on the Mississippi River. Twin 250 hp Mercury outboards and jet drive combine to give the vessel a top speed of 48 mph. A 6-liter Kodiak engine with American Turbine jet power provides more than 3,000 gpm delivered through two Elkhart monitors and three 5-inch outlets.

Small communities sometimes broke from the traditional practice of municipally owned and operated fire-rescue boats, devising and implementing alternative ways to protect and serve their waterways. In Northern Minnesota, Lake Vermilion has over 1,200 miles of shoreline. Since volunteer emergency service often had to travel up to forty miles to reach water emergencies, and then borrow private boats to reach

Since 1975, Virginia's Smith Mountain Lake Marine Volunteer Fire/Rescue Department has operated a fleet of fireboats to serve the 20,000-acre lake community. All are specially built, pump-equipped Boston Whaler variants, such as this 1985 boat rated at 500 gpm. Courtesy of Curt Elie.

One of six units operated by the Hammondsport, New York, fire department, this 2008 Harbor Guard Firehawk Model 26 is powered by twin 225 hp outboards, with 300 hp jet drive to pump 1,600 gpm. During the 1960s, a military surplus DUKW served as the emergency response vessel on Keuka Lake. Courtesy of Hammondsport Fire Department.

The St. Paul Fire Department acquired this 31-foot, 3,000-gpm vessel from Viking Welding & Fabrication. Grab rails for people in the water are recessed into the hull. Author's Collection.

Hall County Fire-Rescue operated this 21-foot Yamaha SRT on Georgia's Lake Sidney Lanier until it was replaced in 2015 by a 23-foot Munson. Costly marina fires, thousands of visitors annually, and 600 miles of shoreline prompted Hall County officials in 2015 to prepare for a boat with firefighting capability. Courtesy of Hall County Fire-Rescue.

A Kevlar hull allows members of the Oshkosh Fire Department to run this Husky Nattig airboat on icy rivers and lakes during Wisconsin winters. Author's Collection.

Nevada's Henderson Fire Department operates this Harbor Guard Firehawk 26 with 500-gpm pump and diesel-jet drive. Author's Collection.

In early 2012, the all-volunteer Culver Union Township Fire Department took delivery of this Neoteric Hovertrek Model hovercraft. Intended mainly for ice rescue on Indiana's Lake Maxinkuckee, the 65 hp craft seats four. Courtesy of Frank Wegloski.

The Grapevine Fire Department of Texas uses this 28-foot SeaArk Little Giant for fire and rescue on Lake Grapevine. Named *Mustang*, the aluminum craft pumps 500 gpm and is powered by twin 225 hp Mercury Verados engines. Courtesy of SeaArk Marine, Inc.

Indiana's Perry-Clear Creek Fire Protection District uses this 2004 750-gpm Scorpion fire-rescue boat on Monroe Lake. Courtesy of Frank Wegloski.

The Clarksville, Indiana, fire department uses this vessel for fire protection and EMS on portions of the Ohio River within its jurisdiction. Courtesy of Frank Wegloski.

There are few limits to the types of equipment fire departments use for fire and rescue on lakes in their jurisdiction. Popular applications of airboats built by American Airboat Corp. of Orange, Texas, include protective services, commercial, and military. The firm's Air Ranger is typically powered by a Chevy 350-cubic-inch, small block engine modified for marine use to drive a pusher prop. Named *Puddle Jumper*, this Air Ranger serves the Port of Sellersburg, Indiana, volunteer fire department. Courtesy of Frank Wegloski.

The Reading, Pennsylvania, Fire Department Underwater Rescue Recovery Squad operated this 1964 Consolidated LARC obtained from military surplus. A glass enclosure was later added to the forward cabin area of the amphibian, named *Charcoal Charlie*. Courtesy of Louis G. Kelter Jr.

In June 2011, fire-rescue boats gather at the Arkansas Fireboat School at Lake DeGray. At left is the *Gist Wilbur* of the Lake Maumelle Volunteer Fire Department. Author's Collection.

the site, residents in 2008 formed a nonprofit corporation to purchase a fireboat. Managed and run by volunteers, the Lake Vermilion Fire Brigade works with various local and government organizations for professional counsel. Fundraising made possible the purchase of a 33-foot Lake Assault fire-rescue craft rated at 1,600 gpm and fully equipped for fire, rescue, and EMS. Built with a landing craft bow and capable of speeds exceeding 50 mph, *Brigade I* is a duplicate of Chicago's Fast Rescue Boat 6-8-8. A 26-foot Boston Whaler named *Brigade II* was donated in 2009, and in 2012, the Brigade received a 25-foot Lake Assault airboat it named *Brigade III;* boats *II* and *III* are equipped with portable pumps.

This fire-rescue boat of the Lake Vermilion Fire Brigade in northern Minnesota is a duplicate of Chicago's vessel number 6-8-8. Built by Lake Assault Boats, *Brigade I* is powered by twin 300 hp Suzuki outboards, allowing speeds more than 50 mph. A 300 hp pump gives *Brigade I* a rating of 1,600 gpm. Courtesy of Lake Vermilion Fire Brigade.

Standards vary for licensing and training boat crews, and agencies using watercraft for emergencies use various methods to certify crews and maintain proficiency. Unique among these communities is the Arkansas Fireboat School, the nation's largest inland water rescue training event. The school conducts annual training for rural fire departments that operate fireboats within the state. Training is provided by the Arkansas Forestry Commission, which, under authority of the US Forest Service, loans boats to rural fire departments that serve populations of 10,000 or less.

Vessels vary in type and propulsion systems, ranging from 20 feet to 50 feet long. Departments that qualify not only are required to attend the school, but must respond to all fires within five miles, equip the vessel with firefighting equipment, maintain liability, and implement an operating policy. Usually, about eighteen volunteer departments attend the annual school, all of which operate one fireboat, with the exception of the Joplin Volunteer Fire Department, which operates four.

Brigade III of the Lake Vermilion Fire Brigade is a 2012 Lake Assault Boats 25-foot airboat powered by a 550 hp Cadillac engine driving twin eight-blade, counter-rotating propellers. Like *Brigade I*, the boat is capable of speeds exceeding 50 mph. Courtesy of Lake Vermilion Fire Brigade.

The amidships cabinet of this 23-foot SeaArk Roustabout houses a Chevy V-6 engine, which also powers a 1,000-gpm pump. The Lakemont-Wiley Volunteer Fire Department operates the customized *Miss Rabun* on Georgia's Lake Rabun. Courtesy of SeaArk Marine.

Since Arizona's Lake Havasu, a reservoir of the Colorado River covering more than 19,000 acres, is a recreational mecca, the Lake Havasu Fire Department provides marine safety. Powered by twin 225 hp outboards and equipped with dual Hale fire pumps, this 2007 30-foot Munson boat proves capable of providing fire protection and life safety. A shallow draft and lowering bow ramp facilitate shore operations. Courtesy of Munson Boats.

CHAPTER 7: CALIFORNIA'S PACIFIC COAST

Los Angeles

After California was admitted to the Union as the thirty-first state in September 1850, interest in developing West Coast ports increased dramatically. Five months prior, Los Angeles had been incorporated as an American city. The rapid growth and industrial expansion of Los Angeles beginning in the 1870s challenged San Francisco's status as the state's major port of entry. Since Los Angeles lacked a modern harbor, a channel was excavated out of San Pedro Bay leading to Wilmington; the channel was then expanded into a harbor. Combined with the Port of Long Beach, the Port of Los Angeles formed the fifth-busiest port facility in the world.

The Los Angeles Fire Department (LAFD) became a paid department in 1886. When the harbor cities of San Pedro and Wilmington were annexed in 1909, the LAFD took on the added responsibility of protection for those ports and regions inland. Beginning in 1911, their eight miles of waterfront were covered by monthly rental of two privately owned fireboats—the *Falcon* and *Warrior*—which were owned by Wilmington Transportation Co. but manned by land-based fire companies. The *Warrior* ended service in 1915, and the *Falcon* in 1917, when the city purchased a barge. Moored at San Pedro beginning in April, the barge accommodated two steam pumpers: a second-size 700-gpm Nott and an Amoskeag rated at 400 gpm.

The Los Angeles Fire Department's early attempts to provide waterfront fire protection took the form of a barge loaded with two steamers, a 1904 Nott and an 1887 Amoskeag. Seen here in 1917 when the makeshift arrangement was new, the barge is loaded with bags of coal for the steamers. Author's Collection.

The first fireboat owned and operated by the Los Angeles Fire Department was the 25-foot *Aeolian* equipped with a 60-gallon soda-acid tank with reel of 300 feet of rubber hose. Author's Collection.

Both were pulled to the barge from Station 36 with two-man crews and the barge was then towed to fires. The department had purchased its first boat, the *Aeolian*, a 25-foot chemical engine vessel, in 1915 to begin service the following year. Chief Archie J. Eley, firm in his belief that these amounted to little more than patronizing by city hall, stated in the 1917 Annual Report in *Fire Engineering*:

> I MAKE THE FOLLOWING RECOMMENDATION THAT A MODERN FIRE BOAT WITH A CAPACITY OF NOT LESS THAN 9000 GALLONS PER MINUTE BE PURCHASED, WITHOUT DELAY . . . I WOULD, ALSO, RECOMMEND THAT THIS BOAT BE CONSTRUCTED OF STEEL, THAT THE ENGINES BE OF THE OIL BURNING, INTERNAL COMBUSTION TYPE, USING THE SAME POWER THAT PROPELS THE BOAT TO ALSO DRIVE THE PUMPS, THEREBY GIVING A LIGHTER DRAUGHT TO THE BOAT AND GREATER ECONOMY IN OPERATION AND MAINTENANCE.

Use of the Aeolian was discontinued in 1919 when the LAFD's first fireboat arrived, although far from meeting Eley's expectations. Built by Marine Equipment Company of Los Angeles, the 64-foot wooden-hulled vessel was powered by four gasoline engines and two 1,250-gpm pumps. It was designated Boat 1 and fitted with a pilothouse a few years later. During the 1930s, it was repowered with new Hall-Scott gasoline engines and two pumps rated at 2,000 gpm each. Not until three years before the boat was taken out of service, in 1968, was it officially christened, becoming the *Archibald J. Eley*. The *Eley* would be one of three LAFD fireboats christened in May 1965, while retaining boat assignment numbers, along with department shop numbers; the *Eley* was given shop number 28.

Although the rapid growth of Los Angeles made it one of the world's largest importers of lumber and a burgeoning petrochemical center, its busy harbor and waterfront lacked adequate fire protection. The fire department's never-ending quest to obtain a powerful fireboat was taken up by Ralph J. Scott when he was appointed chief engineer in 1919. Known for his fire service innovations, Scott's persistence paid off in May 1924 when he gained approval for a fireboat, along with quarters and associated land apparatus. After reviewing plans by ship designer L. E. Caverly, Scott noted in the 1925 Annual Report, "I do not hesitate to say that it will represent the last word in fire boat construction, and that it will fill a most vital

Los Angeles Fireboat No. 1 in 1919. The low-profile, gasoline-powered boat eventually was rebuilt with a pilothouse and named the *Archibald J. Eley*. On the forward deck is a fake, which is rope carefully coiled to prevent entanglement when run out. Author's Collection.

Abaft of the *Eley*'s pilothouse was a ventilator for the engine room, and a hose reel for 2½-inch hose. On the deck were three monitors and eight discharge outlets. Courtesy of Larry Ford Collection.

Los Angeles' Boat 2 on the ways at Los Angeles Shipbuilding & Dry Dock Co. in 1925. The placard below the bow identified it as the shipyard's hull number 47. Courtesy of Los Angeles Fire Department.

need in the protection of that important district." Los Angeles's new fireboat may not have been the "last word" in fireboat design, it was, nevertheless, unique.

The contract for the steel-riveted boat went to Los Angeles Shipbuilding & Dry Dock Co. (later Todd Pacific Shipbuilding) of San Pedro. The shipyard's hull number 47 was launched on October 20, 1925, it was christened *Los Angeles City No.2* by Chief Scott's wife on the twenty-fifth, it went in service on November 27, and was officially commissioned on December 2. Displacing 152 tons, it measured 99 feet long, with a beam of 19 feet. In its engine room were five Winton six-cylinder 350 hp gasoline engines, and two Hall-Scott 12-cylinder 640 hp engines, which drove six Byron Jackson, four-stage centrifugal pumps that delivered a total of 13,500 gpm. The boat's unique triple screw system used a 300 hp Winton engine to drive the main 27-inch diameter propeller, while the Hall-Scotts drove both 48-inch-diameter Wing props. In addition, a pair of 50 hp four-cylinder Hall-Scotts powered two electric generators. More than 2,100 gallons of gasoline held in two tanks allowed long on-station time. *Los Angeles City No. 2* was considered the fastest large fireboat afloat, topping out at 20 mph. Pumps fed five monitors, one of which, mounted atop the pilothouse, was a giant appliance nicknamed Big Bertha. Able to accommodate a 6-inch-diameter tip, at 200 psi, Big Bertha could deliver 10,200 gpm producing a stream reaching 400 feet. Also unique was a monitor mounted atop a tower pipeline that elevated 12 feet, placing the monitor 42 feet above the waterline, a feature adopted for other fireboats. Perforated pipelines lined the deckhouse and upper hull to create a spray protecting boat and crew from intense heat. At deck-level were twenty-four discharge outlets for large-diameter hose, with 3,000 feet of hose carried on four reels. For blazes at petrochemical sites, 300 gallons of foamite was carried, which, when expanded with water, generated 3,000 gallons of foam. Included among Boat 2's extensive inventory of firefighting equipment was a unique floating monitor for fighting under-wharf fires at high tide.

Enhancing the fireboat's originality was Station 112's quarters that incorporated a massive covered boathouse, complete with a boat tender and foam-carrying apparatus; a second boat tender was stationed at Engine 53.

Since diesel power had yet to gain wide acceptance, the LAFD held its preference to gasoline power, despite the danger of explosive vapors. That, however, necessitated a rather complex exhaust system that exchanged the air in the engine room every five minutes. Additional safeguards took the form of eighteen 50-pound carbon dioxide cylinders on deck, in manifold array for piping into the engine room. Two hundred feet of metallic hose allowed the system's use on ship fires. Boat 2, as it was commonly called, operated with a crew of fourteen that comprised captain, pilot, mate, three engineers, an auto fireman, and seven firefighters; the auto fireman, who operated all apparatus without a pump, likely was assigned to boat tenders.

Displaying ten powerful streams, Fireboat 2 officially opens the 11th Mid-Winter Regatta of Los Angeles Harbor on February 20, 1937. Courtesy of ACME News.

Three months after it was commissioned, on March 3, 1926, *Los Angeles City No. 2* and crew fought their first major blaze when the fully loaded lumber schooner *Sierra* burned at wharfside at San Pedro's E. K. Lumber Company.

To better utilize its two fireboats, the LAFD contracted with local shipbuilder Fellows & Stewart to build a cruiser to serve as a command platform and fireboat for yachts and smaller craft. The wooden boat measured 38 feet long, with a 10-foot beam. It was powered by a pair of 250 hp Sterling Marine gasoline engines. Christened *Los Angeles City No. 3* and assigned shop number 870, Boat 3 was launched on March 15, 1928. Boat 3's firefighting capability was derived from a Byron Jackson 290-gpm centrifugal pump to supply three discharge outlets, along with a foam generator. Also aboard were 500 feet of hose and two 50-pound carbon dioxide cylinders. Quartered with Boat 2 and assigned a crew of two, its size, speed, and shallow draft allowed access to all parts of the harbor.

Seventeen streams are visible as *L.A. City No. 2* salutes the opening of National Foreign Trade Week in May 1938. The tower mast is extended and rail standees along the bulwarks allowed nozzles to be secured. Author's Collection.

In the constant fight against age and saltwater deterioration, and to extend its life, Boat 2 underwent modernization in 1945. Painted low-visibility gray for wartime, Boat 2's Winton engines were replaced by five 275 hp six-cylinder Hall-Scotts and two 625 hp V-12 Hall-Scotts; the work was done by Crown Firecoach, a prime builder of apparatus for the LAFD. That same year, Boat 3's engines were replaced by 275 hp Hall-Scott engines.

To replace the 1925 boat tenders, in 1952, the LAFD took delivery of two boat tenders built by Mack and assigned shop numbers 716 and 717, the latter designated Boat Tender 36.

With the planned expansion of Los Angeles Harbor and subsequent rise in boat traffic, the LAFD kept

The fireboat L.A. City No.2, later named the *Ralph J. Scott*, nestled into its unique, cavernous boathouse. Later called Station 12, the structure was a Los Angeles landmark until it was demolished in 1986 to make way for a cargo container complex. Courtesy of Curt Elie Collection.

LAFD Boat Tenders 1 and 2 were special-built 1925 Macks that carried 3½-inch hose to take Boat 2's firefighting punch farther inland. Both tenders featured large fireboat-type monitors. Boat Tender 1 was assigned to Engine 53 and Boat Tender 2 at Station 112, berth 227. Courtesy of Los Angeles Fire Department.

The fireboat L.A. City No. 4 puts on a spectacular display with 16 streams. This view lends credence to its nickname, "the Great White Swan." Courtesy of Curt Elie Collection.

In addition to monitors, rail standees were used to take full advantage of fireboats with ample pumping power. Such was the case with L.A. City No. 4, which could deliver 9,000 gpm. Courtesy of Curt Elie Collection.

pace by adding a fourth fireboat. Boat 4 resulted from research begun in 1954 and was considered the most complete and complex fireboat designed, incorporating many features of land-based fire apparatus. Measuring 76½ feet from stem to stern, and wide in the waist at 24 feet, the all-steel vessel was designed with the firefighter in mind and the ability to handle a wide range of emergencies. Versatility was the byword as the boat took shape on the drawing boards of L. C. Norgaard & Associates of San Francisco. In 1961, the contract for construction of *L.A. City No. 4* was let to Albina Engine & Machine Works of Portland, Oregon.

Largely responsible for Boat 4's unique design was LAFD Battalion Chief Bethel F. Gifford, who commanded marine operations and oversaw the boat's development. Gifford's intent that the boat reflect marine firefighter input from the keel up is obvious in his words in *Firemen's Grapevine* (1960):

> Fire boats, in general, are designed by Naval Architects who are usually hampered by a lack of firefighting experience, by a lack of detailed specifications written by experienced firemen such as are furnished to manufacturers of land based fire apparatus, and by precedent which results in designing a boat first and then adapting the firefighting machinery to the boat . . . Descriptions of fire boats usually stress only three features, the boat itself, the large capacity of fire pumps, and the large turrets. Mention is seldom made of any provision for efficiently fighting medium or small fires which constitute probably 95% of all fires, or large fires under wharves where properly located small streams are more effective than streams from the large turrets even if they could be directed on the fire.

Members of the Los Angeles Fire Department were impressed when Boat 4 was delivered on February 22, 1962. Regardless of the nicknames bestowed on it—technically, it was shop number 60092, while to others, it was the Great White Swan—there was no doubt that it was built for work and up to any challenge. Beneath Boat 4's deck were six 400 hp Cummins diesel engines—two driving twin controllable-pitch propellers and four powering two 3,000-gpm De Laval two-stage centrifugal pumps—and two 1,500-gpm auxiliary pumps. A separate 1,500-gpm pump was used to power four maneuvering jets in the hull.

Besides the boat's five monitors, six 500-gpm rail pipes were fixed at bulwarks on both sides of the vessel for nozzle attachments. This unique piping design replaced rail standees and kept decks clear of hose. In addition, 2,500-gpm under-wharf nozzles were installed port and starboard in the forward hull. Innovative movable hose beds, which resembled those used on land apparatus, replaced unwieldy reels commonly seen on fireboats decks. A 550-gallon foam tank was connected by piping to outlets at deck level to deliver 65,000 gallons of foam. After two weeks of training, *L.A. City No. 4* was placed in service with a crew of five on April 9, 1962.

May 8, 1965, in California marked Fire Service Day, when Los Angeles Fire Department Fireboats 1, 2, and 4 finally were christened in honor of chiefs who played a key role in their design and procurement. Boat 1 was renamed *Archibald J. Eley*, Boat 2 became *Ralph J. Scott*, and affixed in gold letters to the bow of Boat 4 was the name *Bethel F. Gifford.* A spectacular water display for thousands of spectators followed the triple christening.

The decade also marked dramatic growth of harbor facilities, increased cargo ship and tanker traffic, a proliferation of hazardous materials, and growing marinas. Meeting the requirements for harbor-wide protection called for a plan by city and fire department leaders, which resulted in sweeping changes and a new look in the fireboat fleet.

At the beginning of the decade, in March 1960, a massive fire that involved the Matson Lines wharf sparked interest in reviving the use of floating nozzles first tested by the Coast Guard during WWII. The Port of Los Angeles waterfront was notorious for nearly thirty miles of wharfs formed by mazes of heavy creosote or oil-impregnated pilings and timbers, topped with 4-inch planks that were paved with asphalt or concrete. The heavy, complex timber

Boat 4, resplendent in new red and white paint scheme, after being named the *Bethel F. Gifford*. Nearly 2,000 feet of hose was carried in the two large boxes on the afterdeck. Courtesy of Chuck Madderom.

The trio of Drake Craft boats purchased by the LAFD in 1967 featured flying bridges, allowing control from above or in the pilot-house. Boat 5 was assigned to Station 110. Courtesy of Larry Ford.

During the 1950s, a Los Angeles marine firefighter works the wheel that controls traverse of Boat 2's bow monitor. Author's Collection.

Los Angeles fireboat crews found anti-slip deck mats not only more efficient than non-skid mastic, but more forgiving in falls. Boat 2's radar mast was added in 1956. Alongside in 1962 is the tender L.A. City No. 3. In the background is *Hawaiian Citizen*, a container ship built in 1928 that was converted from the WWII attack transport USS *Goodhue* (APA-107). Author's Collection.

construction of Matson's wharf allowed the spread of a stubborn blaze that defied the best efforts of firefighters, giving rise to the idea of SCUBA-equipped firefighters taking floating nozzles under piers. Captain Warner L. Lawrence spearheaded the design and development of this unique equipment. Assistant chief William W. Johnston Jr. took up the cause and developed a SCUBA firefighting unit. New under-wharf tactics were developed that placed divers aboard fireboats responding to dock fires. Gripping floats that incorporate nozzles connected to boat hoses, divers use nozzle back pressure for thrust and maneuverability. After fireboat and land streams are shut down, they move in to attack hard-to-reach flames. The bright yellow floats can be interconnected, supplied by a single 2½-inch hose, and be propelled beneath a dock, forming a powerful water curtain. This innovative concept, while often viewed with skepticism, led to similar operations in other major cities, such as Boston and Tacoma.

Another major change, which would better use the fire department's SCUBA team, was the decommissioning of Fireboat 3 in July 1967, and the acquisition of three 34-foot fireboats from Drake Craft Corp. of Oxnard, California. The fiberglass-constructed boats were designed for fire-rescue in marinas, and as SCUBA platforms and floating command posts. The boats were capable of 30 mph thanks to a pair of Chrysler 290 hp marine outboard engines. A center-mounted engine drove a Waterous 750-gpm pump to supply a Stang bow monitor. Since the pumps, plumbing, and monitors were installed by Crown Firecoach, the boats often were called Crown-Drake Crafts. Crown's job designation, "CBSW 34-750," stood for Crown Boat Salt Water, 34-foot, Centrifugal Pump 750 gpm. Included in the versatile craft was a 50-gallon foam tank.

Boats 1 and 3 were built in 1967 and assigned to Stations 111 and 49, respectively, while Boat 5 was completed in 1968 and assigned to Station 110; all were operated with a crew of three, one of them a mate to pilot the craft.

Further modernization of the fleet included major rebuilding of the *Ralph J. Scott*, which began at San Pedro's Fellows and Stewart yard during early 1969. Cost-conscious chief engineer Raymond M. Hill had all but written off the *Scott* when Captain Lawrence, a skipper of the vessel for twenty-eight years, made his convincing pitch for updating the boat at far less cost than a new boat. Significant among the changes was the addition of a 50-foot Snorkel articulating platform mounted next to the tower. Success with the *Gifford's* under-wharf nozzles led to their installation in the *Scott,* along with new hydraulically operated Stang monitors. Maneuvering jets, along with hydraulic steering and pilot control of engine, completed the update. The *Scott* resumed operation in late October 1969 with its crew reduced from fourteen to eight, thanks to the overall efficiency afforded by the upgrades.

Throughout its long career, the *Scott's* crew braved horrific, dangerous blazes, including spectacular fires in 1944, 1947, and 1975, all of which involved flammable liquids. Repeatedly, they took their trusty fireboat into infernos that were far beyond the reach and firepower of land companies. But even the mighty fireboat and crew was no match for the protection demanded by massive petrochemical facilities and the switch from break-bulk freighters to giant container ships. The Los Angeles Fire Department needed more firepower on the water and overall significant improvements in fireboat operation. Plans initially proposed by seasoned fireboat pilot Warner L. Lawrence were given a hard look and put into play beginning in 1975. Most notable in the three-year upgrade was the switch from gasoline power to diesel. The *Scott's* new power system comprised three 380 hp six-cylinder Cummins diesels, two 525 hp Detroit V-12s, and a pair of 700 hp Cummins V-12s. The switch to diesel power boosted the *Scott's* gpm water delivery to more than 18,000 gpm at 150 psi. "Big Bertha" alone could deliver a 10,000-gpm knockout punch, its stream

A crewman secures a hawser on the capstan during mooring of Fireboat No. 2 in 1967. The hand-operated bow monitor is tipped with a nozzle that adjusts from straight stream to fog. Fastened to the bulwark is a spare monitor with pipeway and long operating handle. Author's Collection.

The *Ralph J. Scott* shows off its new look after rebuild in 1969. Huge monitors are remotely controlled Stangs, while a 50-foot Snorkel boom was added next to the tower, which is fully raised. Author's Collection.

The 1925 vintage *Ralph J. Scott* doesn't show its age in this early 1970s view. During the mid-1970s, it again underwent refurbishment to extend its career, which included the switch to diesel power. Courtesy of Chuck Madderom.

Aboard the *Scott* on April 12, 2003, at LAFD Station 112, then-Councilwoman Janice Hahn christens the new Boat 2. The transitional event also marked the simultaneous retirement and replacement of Fireboats 1, 2, 3, and 5, and their christening. Courtesy of Donald Root.

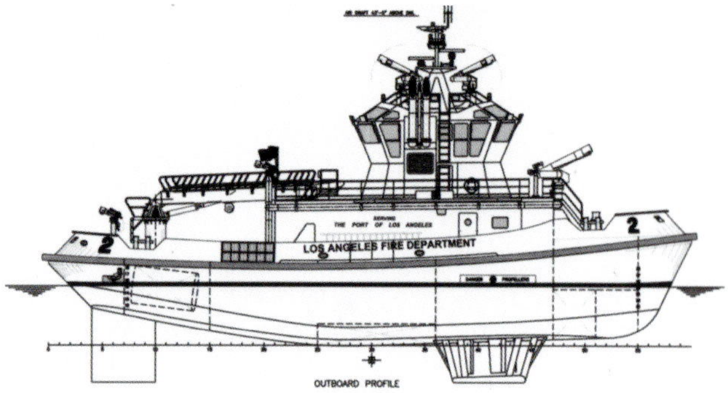

The outboard profile of the *Warner L. Lawrence* reveals the deep hull and extreme downward projection of the cycloidal drive system and associated skeg. Author's Collection.

reaching 500 feet. The change also meant the elimination of the bank of bulky carbon dioxide cylinders. The *Scott's* six Byron Jackson bronze pumps remained unchanged, having given reliable service for more than half a century. Despite these improvements, the Los Angeles Fire Department had to make a giant leap to stay on equal footing with the port's immense growth.

At the dawning of the twenty-first century, the Port of Los Angeles covered 8,000 acres, including forty-three miles of waterfront used for commercial, industrial, and recreational interests. As the chief container port in the US and a major relay center of international cargo, the port is a gateway to the world. Pier 400 alone, covering 600 acres, is the largest container terminal in the world. It becomes clear, then, how a city in economic strife justified spending nearly $12 million on new fireboats. Former LAFD assistant chief John D. Badgett relied on his experience in harbor operations to pen an article for the LAFD newsletter, *Fire Watch*, in which he puts in perspective the acquisition and operation of modern, highly capable fireboats to maintain the port's reputation as the "World's Safest Harbor." Badgett explains that the new boats are the result of a ten-year joint effort between fire department and port officials, beginning with a needs assessment and identifying all hazards. Unsurprisingly, intense research pointed to the top

The L.A. Fireboat 2 is a study in modern technology applied to firefighting vessels. Besides a full EMS suite, telescoping boom, and cycloidal drive system, the *Lawrence* features a dive platform that retracts ten feet below the surface. Courtesy of Chuck Madderom.

priorities in a new fleet of fireboats as strength, longevity, technological superiority, and multi-function.

Replacing the venerable *Ralph J. Scott* would be new Fireboat 2, designed by Robert Allan LTD of Vancouver, British Columbia. In view of their experience building tractor work boats, Nichols Brothers Boat builders on Washington's Whidbey Island was selected for construction. Like its predecessor, the new boat stretched 105 feet, although it was wider in the beam at 29 feet. Constructed of steel with an aluminum superstructure, it is powered by four 1,800 hp Detroit diesel engines shafted to a Voith Schneider cycloidal drive system allowing a top speed of 16 mph. The cycloidal system and associated stern skeg for directional stability gives the boat a deep draft of 15 feet. Engines power four Unitor 7,500-gpm pumps, and a pair of 1,575 hp diesels power 7,000 and 4,500-gpm pumps. When hull number S-143 was launched on Whidbey Island on January 17, 2003, it was the most powerful fireboat afloat, boasting a total pump rating of 38,000 gpm. It sports ten monitors: a bow monitor capable of delivering 11,000 gpm, six monitors rated at 1,000, 3,500, and 5,500 gpm, plus two 2,000-gpm under-wharf monitors at the stern. Pumps also supply eighteen discharge outlets, along with a pipe-way incorporated into a 50-foot, 1,000-pound capable telescoping boom by Alaska Crane. Monitors are foam capable, drawing from a 6,000-gallon foam supply.

The innovative, high-tech vessel arrived at San Pedro on April 21, and on May 12 it was appropriately christened *Warner L. Lawrence* in a ceremony that retired the seventy-eight-year old *Ralph J. Scott*. The *Lawrence* began service with a crew of eight, which often is supplemented with firefighters and LAFD divers, who continue to play an important role in harbor protection.

The ceremony that dedicated the *Lawrence* and ushered out the *Scott* also marked the retirement of Boats 1, 3 and 5, and introduced their replacements. The trio was designed by Jensen Maritime Consultants and built during 2002 by Kvichak Marine Industries of Seattle. All are identical, measuring 39 feet long with a 13½-foot beam. Constructed of aluminum, the boats are powered by a pair of 450 hp Caterpillar diesels, giving them a top speed of 34 mph. A Cummins diesel drives a 2,400 gpm Hale pump to supply a 1,000-gpm Stang bow monitor and six discharge outlets. A 50-gallon foam tank is included.

The four new boats, along with Boat 4, the *Gifford*, the oldest of the fleet, provide the most technologically advanced protection to the Port of Los Angeles and surrounding communities. After celebrating a half century of service, the *Gifford* eventually was sold. Its new name, *American Triumph*, seems appropriate in reflecting on its career.

On a broader scope, the Los Angeles County Fire Department (LACFD) covers 2,305 square miles with 171 stations, making it the fourth-busiest fire department in the nation. Although marine operations are largely the responsibility of the LAFD and other waterfront communities, the LACFD operates a fleet of ten pump-equipped Baywatch rescue boats, along with Boat 110, a front line vessel at Marina Del Rey. In 2011, the department received from the US Coast Guard a 41-foot utility boat that serves as Reserve Boat 310. The Coast Guard used the UTB for a variety of missions in moderate weather and sea conditions where its speed and maneuverability proved ideal.

Boat 110 is similar to the LAFD's 39-foot boats, having also been designed by Jensen Marine and built by Kvichak Marine Industries. Power is derived from twin Caterpillar diesels, which give a top speed of 30 mph. A Cummins diesel drives a 2,400-gpm Hale pump to supply a 1,000-gpm Stang monitor.

Long Beach

A major container port, Long Beach is second in size only to the neighboring Port of Los Angeles. The Marine Safety Division of the Long Beach Fire Department is responsible for fire protection of the city's twenty-five miles of shoreline, which incorporates 5,000 acres of oceanfront. Waterborne fire protection for the port began in 1942 with the 30-foot wooden fireboat *Chas. H. Windham*, named for former Long Beach mayor Charles Windham, who was committed to development of the harbor. Rapid expansion of

The first fireboat of the Long Beach Fire Department was this 30-footer named *Chas. H. Windham*. During his terms as mayor early in the twentieth century, Windham was committed to development of Long Beach Harbor. Equipped with a bow monitor, the wooden boat began service in 1942. Courtesy of Curt Elie Collection.

Long Beach Fire Department's Fireboat 1, along with its twin, arrived from nearby Fellows and Stewart Shipyard in 1954. The 4,500-gpm fireboat is seen here in 1982 near the legendary aircraft *Spruce Goose*, which was displayed in the harbor that year. Courtesy of Larry Ford.

Intended as replacements for Long Beach's twin 40-foot fireboats, *Challenger*, and its twin, *Liberty*, within their first two years of service were found riddled with design and construction deficiencies, and showing signs of corrosion. The 10,000-gpm *Challenger* is seen here shortly after delivery in 1986 in its original scheme and without bulwarks. The crane on the stern was rated at 1,000 pounds. Courtesy of Curt Elie Collection.

the port led to the acquisition of two identical 40-foot steel fireboats rated at 4,500 gpm. The first, Fireboat 1, was launched dockside at Fellows and Stewart Shipyard on May 4, 1954, while its twin, Fireboat 2, was launched on June 22. Located in Los Angeles Harbor, the popular shipyard opened its doors in 1896, remaining in business for more than seventy years. Later, the twins would be re-designated Fireboat 15 and Fireboat 20, respectively. Under reciprocal agreement with the Los Angeles Fire Department, the two vessels and LAFD's fleet provided maximum effort at major blazes in the port region.

Although satisfied with their two fireboats, Long Beach marine firefighters in 1985 looked forward to a pair of larger, more powerful, state-of-the-art fireboats on the drawing board at Seattle marine engineering firm Nickum & Spaulding Associates. Delivery of the identical diesel-powered, twin screw fireboats was slated for September 1986. Measuring 88 feet long, the boats had steel hulls and aluminum superstructures, and featured fully automated engine rooms that allowed operation of the boats by as few as two crew members. Their 10,000 gpm at 165 psi pumping capability was divided among three pumps—two 2,500-gpm units powered by main propulsion engines, and a single 5,000-gpm unit driven by a dedicated 600 bhp diesel. All pumps supplied a 12-inch-diameter surrounding main that fed seven monitors and three discharge manifolds. Monitors, all of which were foam-capable, included one atop a 64-foot telescoping tower, two under-wharf monitors, and an 8-inch-diameter monitor atop the pilothouse. Design requirements combined twin 760 bhp Detroit diesel engines and V-hull to ensure quick arrival at all port emergencies. Given the number of oil drilling platforms and petrochemical facilities at Long Beach, all monitors were equipped with foam nozzles supplied by a 1,000-gallon tank of foam concentrate.

Both vessels were built by Moss Point Marine of Escatawpa, Mississippi, during 1986 and began service as the *Liberty* and the *Challenger* in 1987. That same year, both design firm and shipbuilder of the pair came under new ownership; Nickum & Spaulding

One of the twin fireboats built by Moss Point Marine for Long Beach undergoes sea trials following rebuild. A troubled maintenance record that began soon after the boats were delivered led to a modification program during the late 1980s. The vessel displays new paint scheme, along with the addition of bulwarks and a monitor tower. Author's Collection.

became Elliott Bay Design Group, and Moss Point Marine operations were assumed by Trinity.

Notes posted by personnel saying, "Sell the red ones, keep the gray ones" were clear indications that elation over the new boats was short-lived. Long Beach fireboat crews and fire department officials alike expressed discontentment over corrosion and maintenance problems that occurred shortly after the boats began service. A growing list of maintenance problems called for costly modifications before the boats reached the two-year mark. When completed, the *Challenger* resumed service at Station 15 and the *Liberty* at Station 20.

Not until well into the twenty-first century would the Long Beach Fire Department reach its maximum potential in fighting fire from the water. In 2011, fire department and port officials collaborated to identify requirements for a new fireboat. Since requirements called for much more than a fireboat, plans on the drawing boards at Robert Allan Ltd. featured a vessel that could provide on-shore support, assist vessels in distress, support dive operations, conduct harbor security, manage EMS, operate in CBRN environments, and serve as command center during disasters. Specifications pointed to a vessel patterned after Los Angeles's *Warner L. Lawrence.* Although similar in appearance to the *Lawrence,* Long Beach's boat would be larger and more powerful, measuring 108 feet long, with a 35-foot beam, and capable of pumping 41,000 gpm.

Since it was intended that both the *Liberty* and the *Challenger* be replaced, the practice of ordering two boats remained in place. The port's immense size and operational tempo, in addition to reciprocal agreement with the Port of Los Angeles, was sufficient justification for two highly advanced, custom-designed fireboats. In late 2012, Foss Maritime was selected to build the vessels at its Seattle yard. From humble beginnings, Foss rose to become a giant in the work boat industry. Founded in 1889, the Foss story is a prime example of the American dream. Thea Foss, a young Norwegian immigrant, turned a used rowboat into a worldwide maritime conglomerate, and the largest tugboat company in the western US. So renowned was Thea Foss that she was the inspiration for the fictional character Tugboat Annie.

Among the fireboat's unique features is its ability to travel at 15 mph and at "low wash" speeds while maintaining maneuverability in tight areas. Low-speed operation is made possible by a Voith Schneider propulsion system driven by two Caterpillar diesel engines. Another pair of like diesels, along with a smaller engine, power fire pumps that supply ten monitors, the largest of which can deliver 12,000 gpm a distance of 600 feet. Two of the monitors can deliver foam at 6,000 gpm a distance of 500 feet. Fore and aft control stations in the pilothouse, plus panoramic visibility, allow the crew of four superior maneuverability and a complete overview of all deck areas and surroundings.

On May 1, 2014, Colleen Espino, wife of Long Beach Fire Department Battalion Chief Xavier Espino, christened the first vessel Fireboat 20 at the Foss yard. As of this writing, its twin, Fireboat 15, is due for delivery by early 2015.

Besides the two fireboats, the Long Beach Fire Department operates three 31-foot rescue boats, each of which is equipped with a 1,000-gpm pump supplying a bow monitor. All were built by Seaway Boat Company of Long Beach, which has been a steady supplier of rescue and patrol boats to California fire and law enforcement agencies.

The Long Beach Fire Department continued its practice of ordering twin fireboats into the twenty-first century. Fireboat 20 was the first of two high-tech 41,000-gpm fireboats with Voith Schneider Propulsion Systems. Noteworthy in this view are the vessel's draft marks, which corresponded to the extreme draft as a result of control blades and skeg of the propulsion system. Author's Collection.

Oakland

Oakland, California, is the largest among more than seventy cities in the eastern region of San Francisco Bay, referred to as East Bay. The significance of Oakland as a military port during WWII nudged it ahead of its rival San Francisco after the latter's nearly century-long reign as principal port on the Pacific. Its growth into a major transportation hub and trade center for the region helped to ramp up the tempo of the port in the postwar era. Although the Oakland Fire Department was well established, fire protection from the water was long overdue. As early as 1909, Alameda Fire Chief F. K. Krauth expressed the need for a fire tug in Oakland Harbor following a serious blaze that destroyed a key bridge. The chief's words went unheard and protection from the water would not come until after the war. Largely through the efforts of Congressman George P. Miller, the famed Pearl Harbor tug USS *Hoga* was transferred on loan to the Port of Oakland for use as a fireboat. The cost of the vessel's operation and crew salaries was to be shared by the Port of Oakland and the City Council.

A navy crew delivered the *Hoga* from Treasure Island to Oakland's Grove Street Pier, where transfer

The harbor tug USS *Hoga* (YT-146) had a storied firefighting career, beginning during the 1941 attack on Pearl Harbor. Here, the tug and crew fight fires burning in the forecastle of the beached and burning battleship USS *Nevada* (BB-36) on that fateful day. Courtesy of US Navy.

Following naval service, the famed tug *Hoga* went to Oakland, California, as a fireboat named *Port of Oakland*. Reconditioning in 1948 brought its pump rating up to 10,000 gpm. Six of the seven monitors are visible, along with three pumps installed on the afterdeck. The center monitor atop the deckhouse was later removed. Courtesy of Rich Schneider.

Although renamed *City of Oakland* after the port ceased financial responsibility of the vessel, the fireboat remained relatively unaltered, maintaining its appearance as a navy tug which, in turn, added to its appeal as a museum display. Author's Collection.

papers were signed on May 28, 1948. Before beginning service, the vessel was reconditioned at Oakland's Pacific Drydock and Repair. When built by Consolidated Shipping of Morris Heights, New York, in 1940, the *Hoga* was powered by twin 650 hp McIntosh & Seymour diesel engines driving two Westinghouse generators. These, in turn, powered two 515 hp propulsion motors that turned a single screw to achieve 18 mph. A pair of 2,000-gpm centrifugal pumps, each driven by its own Buda diesel engine, supplied three monitors and six discharge outlets, plus a protective spray system. During reconditioning, three 225 hp Buda diesels and United Iron Works two-stage centrifugal pumps were added to the afterdeck, boosting its pump rating to 10,000 gpm. Additional discharge outlets and four monitors were added, one of which was later removed. The 100-foot *Hoga* was re-named Fireboat 1 and began service on December 1, 1948. On board around the clock were a pilot and marine engineer, with a crew of seven firefighters and a battalion chief of the Oakland Fire Department assigned to a firehouse immediately adjacent to the boat's berth.

The vessel was officially commissioned *Port of Oakland* on December 1, 1949, and the next day fought a fire aboard the freighter *Hawaiian Rancher.* The freighter suddenly listed, snapping its mooring lines, threatening to capsize the fireboat alongside. But the venerable fire tug again was spared and went on to serve the Port of Oakland for nearly five decades. During that time, it remained on loan status, technically still a navy vessel. Accordingly, in February 1962 it was reclassified from *YTB-146* to *District Harbor Tug, Medium YTM-146.* The following year, when the Port of Oakland relinquished responsibility for shared financial support of the *Port of Oakland, it* was renamed *City of Oakland.*

One of the fireboat's lighter moments had President Jimmy Carter aboard for a tour in 1980, during which he playfully turned the bow monitor at an accompanying press boat. On December 7, 1982, the *City of Oakland* was relocated from its long-time Broadway berth to trendy Jack London Square, named after the author and owned by the Port of Oakland. The vessel achieved National Historic Landmark status in June 1989.

As harbor operations changed and the fireboat aged, city officials eyed a smaller, more efficient vessel to protect the modern waterfront. Although the *City of Oakland* could function as a portable pumping station in the event of natural disaster, the 1989 Loma Prieta earthquake and the Oakland Hills firestorm in 1991 drove home the importance of developing a modern method of providing the city with water. Nearby San Francisco had learned this valuable lesson and maintained its emergency water system. Both cities had built auxiliary water supply systems after the 1906 earthquake, with Oakland's covering 370 acres of the downtown area. Although Oakland lacked a fireboat, a waterfront manifold was included. During construction of the Bay Area Rapid Transit system, Oakland's water system was shut down and was never re-opened, much less maintained. Not until the Loma Prieta disaster did city officials recognize the need for a modern emergency water supply system. Oakland citizens overwhelmingly voted to allocate funds to upgrade the city's emergency vehicles, which included over a million dollars for a

Used in conjunction with powerful fireboats, the Oakland Fire Department's Portable Water Supply System (PWSS), consisting of four trucks carrying more than a mile of large-diameter hose, can supply large volumes of water long distances to areas where water supply is disrupted by disaster. Courtesy of Chuck Madderom.

Besides one mile of 5-inch-diameter hose, this 1994 Ford Cargo 8000 of the Oakland Fire Department carries four portable hydrants, various fittings, a monitor, and a dozen hose bridges. Portable hydrants line the tailboard and six Gleeson valves are carried on sideboards. The red appliances are wyes to supply two 3-inch-diameter hose lines. Courtesy of Chuck Madderom.

new fireboat. Plans were laid for the new vessel, and Oakland Fire Department members began work on a Portable Water Supply System (PWSS). From their designs emerged four trucks, each with more than a mile of 5-inch hose to transfer water. Used for fire suppression in conjunction with the fireboat, the four trucks combined can supply emergency water in the event of disaster. Since it was fielded in 1994, the system has been used effectively at several massive fires.

Construction of Oakland's new fireboat got under way in 1994, spelling the end of the venerable *City of Portland. It* was returned to the navy, placed on donation hold due to its historic status, and mothballed in 1996 at nearby Suisun Bay as part of the National Defense Reserve Fleet. After the Arkansas Inland Maritime Museum won the bid in 2005 to acquire the former *Hoga, it* was made seaworthy at Mare Island for the 2012 journey to its new berth on the Arkansas River.

Oakland's new fireboat is an aluminum-hulled, 64-foot vessel built by Sea-Fab, Inc. of Pascagoula, Mississippi. Hull number 156 took eight months to build and left Sea-Fab's yard in fall 1994. After arriving at Oakland, a contest was held to name the boat, with the winner, *Sea Wolf*, taken from one of Jack London's literary works since the boat was berthed next to Jack London Square. The *Sea Wolf* is based on a conventional tugboat, having twin screws and thrusters for maneuverability. Twin 535 hp Caterpillar V-8 direct-drive diesels also power a pair of Aurora centrifugal pumps rated at 3,750 gpm each to supply three monitors.

The Oakland portion of East Bay became more dangerous in 2003 when the city ordered the *Sea Wolf* out of service to save costs. Fire Station 2, which was part of the fireboat contingent, also fell victim to the budget axe, stripping the prime tourist area of fire and EMS protection. Although the nation's fourth-largest port, Oakland's became the only one without a fireboat. The city, however, made known its intent to keep the fireboat operable since it can supply Oakland's above-ground water system in the event of an earthquake.

Helping to fill the fireboat void in Oakland Harbor is nearby Alameda, which in 2012 received a Homeland Security grant for a new fireboat. The contract for the 2,000-gpm vessel went to Harbor Guard Boats for its 32-foot model Firehawk. On February 13, 2014, the boat was christened *Deanna Jo*, after an administrative technician of the Alameda Fire Department. Alameda had been without fireboats since 2008, when the city disposed of two former Coast Guard boats—a 40-foot UTB and a 32-foot PWB, the latter having gone to the Santa Cruz Fire Department in 2012.

San Francisco

During the nineteenth century and well into the twentieth century, San Francisco was hailed as a booming seaport and California's major port of entry. The city's phenomenal growth during this period often was tempered with setbacks, with much of the city destroyed by six major fires. The conflagrations gave rise to volunteer fire companies beginning in 1849, with a municipal fire department begun in 1866. Forty

This 40-foot ex-Coast Guard utility boat was one of two boats operated by the Alameda, California, fire department. Courtesy of Curt Elie Collection.

Even before the City of San Francisco placed dedicated fireboats in service, city officials identified and met the needs of water supply in the event of earthquakes. A new high-pressure battery, which was pulled by a single horse, is being tested in 1903. Two of these batteries were in use when the 1906 earthquake struck. Author's Collection.

years later, the most pivotal event in the fire department's history would occur, which remains one of the worst natural disasters in the nation's history—the Great Earthquake and Fire of 1906. The quake occurred in April, setting off fire storms that burned much of the city for four days, killing thousands of people, with fatality estimates ranging between 3,000 and 6,000.

At the time of the quake, San Francisco had no fireboats; however, a pair of state-owned tugboats with firefighting capability assisted, along with Mare Island fireboats and two US Army fireboats; these boats are credited with saving piers and shoreline structures. The fire tugs, the *Governor Markham* and the *Governor Irwin*, were identical, having been built in 1878. They measured 86 feet, with a 20-foot beam, each having two coal-burning, non-condensing steam engines that powered Hooker piston pumps to deliver 1,100 gpm.

In the aftermath of the tragedy, San Francisco officials wasted little time in identifying improvements necessary to meet future disasters, especially earthquakes and fires. As the fire department struggled to return to full strength, plans for a pair of large fireboats were set in motion. Finding a builder for boat and machinery proved simple within the city's Potrero Point industrial district, which included historic Pier 70. Prominent in the industrial enclave at the pier was Risdon Iron and Locomotive Works, which had cut its teeth building mining dredges, locomotives, and ships since the gold rush.

The identical fireboats each displaced 272 tons and stretched 120 feet, with a 26-foot beam, drafting 12 feet of water. Only New York's *Duane* and *The New Yorker*, and Duluth's *McGonagle* were longer. San Francisco's twin fireboats were intended as floating pumping stations as much as they were firefighters. So mindful were city fathers of the devastation wreaked by the earthquake and fires that they saw the boats only as an auxiliary water supply system, and addressed them as such. Their pumping systems were indeed unique, even compared to those pioneered in New York. Power was derived from two cross-compound vertical condensing steam engines with two Babcock & Wilcox oil-burning boilers. They, in

San Francisco's first dedicated fireboats arrived in 1909. Measuring 120 feet with a sweeping sheer, the 9,000-gpm *Dennis T. Sullivan* was one of a pair intended for use as floating pumping stations as much as for firefighting. Unlike large fireboats of the Midwest that had to negotiate low bridges, West Coast fireboats, such as the *Sullivan*, featured high funnels and a fixed monitor tower, which was enclosed. Author's Collection.

turn, powered a General Electric Curtis steam turbine that drove two multi-stage Byron Jackson centrifugal pump units, each rated at 4,500 gpm at 150 psi. Within each boat's water main loop were three monitors, manifolds incorporating twenty outlets, and a protective spray system. Pumps also supplied standee nozzles mounted along the bulwarks, and two portable batteries with Gorter nozzles for use at six connections about the boat. One monitor was atop a 55-foot telescoping tower. Each boat carried 7,000 feet of hose.

Both vessels arrived in 1909, with the first assuming its watch at Embarcadero—the eastern waterfront of the Port of San Francisco—at the foot of Harrison Street as Fireboat 1. It also bore the name *Dennis T. Sullivan* in honor of the department's fourth chief. Sullivan died as a result of injuries sustained when the 1906 earthquake caused an adjoining building to topple onto the firehouse in which he lived. The second vessel, designated Fireboat 2 and named *David Scannel* after the department's third chief, was berthed at Embarcadero at the foot of Bay Street. A crew of thirteen was assigned to each fireboat, which gave San Franciscans some measure of assurance. But the large, powerful fireboats were only part of the solution to the region's response to natural disasters that were sure to come.

San Francisco's water supply has remained a significant issue throughout the city's history. As early as 1895, Chief Sullivan, in scathing testimony in the wake of numerous large fires, chastised those responsible for inadequate water mains and pressed for more and larger mains. Large fires continually broke out, but it was the disastrous Baldwin Hotel fire of 1898 that further exposed weaknesses of the city's water system. City officials were left with no choice but to heed Sullivan's words and augment the fireboats, not only with cisterns but with eight land-based General Electric turbine units capable of pumping 2,700 gpm of saltwater at 300 psi. Sullivan pushed for a gravity water tank atop Twin Peaks, and street mains designated for firefighting. In a move to lessen reliance on steam engines, Sullivan proposed installing high-pressure hydrants throughout the high-value district, and then using hose carts with manpower and equipment to work from the hydrants. Chief Sullivan truly became the father of modern high-pressure water systems, which remain in use today in various forms.

San Francisco's fireboat fleet got a boost in pumping power when the vessel *Frank G. White* was launched in 1949. Owned by the state and operated by the Port of San Francisco, the *White* would back up the *Scannel* and the *Sullivan.* Although primarily a tugboat, the

The twin to San Francisco's *Dennis T. Sullivan* was the *David Scannel*, seen here in its original white livery with buff-colored funnels and black skeleton tower with Gorter monitor. Author's Collection.

Fresh from the yard of Pacific Coast Engineering Company as hull number 140, the *Frank G. White* undergoes sea trials in 1949. With fire pumps capable of delivering 6,000 gpm, the *White* was as much a fireboat as tug. Courtesy of Curt Elie Collection.

San Francisco's fireboat *Phoenix* battles its first major fire, a four-alarm blaze at the famous San Francisco Ferry Building on April 1, 1955. Author's Collection.

72-foot *White,* under agreement between the San Francisco Fire Department and the port, responded to major waterfront fires, where its 6,000-gpm firefighting capability was put to use. The *White* was built by Pacific Coast Engineering Company of Alameda.

Having served for more than forty years, the *Sullivan* and *Scannel* were proving costly to maintain. Keeping one boat at the ready meant keeping its oil-burning boiler at full pressure, while the boiler of the other boat was kept hot. Not only did this state of readiness consume a great amount of fuel, the boilers had to be constantly monitored. The time had come to update the fleet.

Having drawn up specifications and requirements for a new fireboat, city officials sent out requests for proposals for the new vessel. Plant Brothers Shipyard of Alameda got the job, delivering the vessel in May 1954. Appropriately christened *Phoenix,* the fireboat was commissioned on June 22. Boat and crew did not have to wait long for their baptism of fire, which occurred one month later.

The *Dennis T. Sullivan* stood down when the *Phoenix* arrived, and the *David Scannel* in mid-June—both later were scrapped. The *Frank G. White* remained, to be placed on active manned status whenever the *Phoenix* went into dry dock. It was decommissioned in March 1978 as a cost-saving measure, but went on to serve as the tug *Jericho Chief,* and later as the *Valiant.*

The *Phoenix* measures 89 feet long with a 19-foot beam and drafts 6 feet of water. Beneath its deck are three 550 hp Cummins V-12 diesel engines for propulsion, and for driving three De Laval two-stage centrifugal pumps rated at 3,200 gpm each. These supply four monitors, one of which is atop a mast that extends to 48 feet above the water, plus two 1,000-gpm under-wharf nozzles, a 14-outlet discharge manifold, and a foam system. It also carries 3,000 feet of 3-inch-diameter hose for connecting to dockside manifolds to pump water far inland. These pre-selected docking sites are but one option in San Francisco's extensive and elaborate Auxiliary Water Supply System (AWSS). Although the probability of earthquakes exists in the San Francisco region, the city's AWSS serves as a model for one of the most vital components of any city's fire protection.

Details of the foredeck of the *Phoenix* during the 1970s. Coaming surrounding deck hatches helped prevent accidents and, to a lesser degree, water from entering lower compartments. Author's Collection.

In 1984, the thirty-year-old fireboat *Phoenix* wore this white over black scheme, then popular for fireboats of large cities. Like land apparatus, fireboats were a source of pride, which was reflected in their pristine condition. Author's Collection.

The San Francisco Fire Department has a long history of hose tenders, which form the department's High Pressure Battery. Two single horse-drawn batteries with Gorter monitors were in use with the San Francisco Fire Department at the time of the 1906 earthquake and fire; one was destroyed. Favored as a high-pressure monitor, the Gorter saw wide use with western fire departments, as did Gorter water towers. Both were invented by Henry H. Gorter during the 1880s. Vertical movement of Gorter's nozzle is on a ball and socket, allowing free flow at any angle, while a roller bearing permits horizontal movement for easy control by one firefighter. The High Pressure Battery later comprised special vehicles carrying large-diameter hose laid from fireboats to supply hose tender monitors and other heavy stream appliances. A significant upgrade of the High Pressure Battery during the 1980s resulted in the PWSS, which was designed to bypass underground mains damaged during an earthquake. The above-ground system combines unique department-made portable hydrants with the tender's 5,000 feet of 5-inch-diameter hose to supply water great distances from sources such as fireboats, cisterns, special hydrants, and open bodies of water. The timing could not have been better. When a major earthquake struck San Francisco on October 17, 1989, the fire department's PWSS and the fireboat *Phoenix* not only were put to the test, they exceeded all expectations.

Named the Loma Prieta earthquake, since its epicenter was near Loma Prieta Mountain, the highest peak in the Santa Cruz Mountains, the quake caused twenty-two structure fires and more than 500 emergency incidents, such as gas explosions, building collapses, and entrapments, over the following seven hours.

Hose Tender 5, seen here in 1967, was a 1946 American LaFrance powered by a 240 hp V-12 engine. Courtesy of Chuck Madderom.

Most pressing was a building fire in the Marina district that threatened other structures. Although the hydrant tapped by the first arriving engine company was on the city's AWSS, the burning building collapsed onto the hydrant. It got worse. Leaks occurred in both water main systems, forcing firefighters to reposition and stretch hose great distances to alternate water sources. More explosions caused buildings to collapse onto those lines. The

Hose tenders of the San Francisco Fire Department's High Pressure Battery were specialized units carrying large-diameter hose to carry water inland from fireboats. The signature ball and socket of Gorter monitors was evident on these hose tenders. The unique device allowed ease of operation by one firefighter. Hose Tender 1 was a 1937 American LaFrance. Courtesy of Chuck Madderom.

The body of Hose Tender 15 was fabricated in the San Francisco Fire Department shop and mated with a 1950 Mack Model 21-LS Type 95 cab and chassis. Hose Tender 15 was powered by a 275 hp, six-cylinder Hall-Scott engine, and it carried 2,000 feet of 3-inch-diameter hose. The yellow wooden blocks are hose bridges. Courtesy of Chuck Madderom.

Hose Tender 8 carried 5,000 feet of 5-inch-diameter hose in a Welch-built body mated with a 1987 Ford. Power to handle the heavy load came from a 275 hp, eight-cylinder Caterpillar diesel engine. The fire department's High Pressure Battery evolved into the Portable Water Supply System. Courtesy of Chuck Madderom.

fireboat *Phoenix* arrived, along with the fire department's PWSS trucks. The fire had progressed to a point where the entire district was threatened, but when fire crews put the *Phoenix* into action supplying two PWSS hose tenders, an engine company and two ladder trucks, the Marina district was saved. The *Phoenix* pumped at more than 6,400 gpm continuously for fifteen hours, delivering approximately 5½ million gallons of water, until the blaze was brought under control.

Not wanting to face the unthinkable if another disaster occurred while the *Phoenix* was laid up or in for its annual maintenance haul-out, it was decided to acquire a second fireboat. In view of the *Phoenix's* faithful service, there was no opposition. However, funds for another fireboat were unavailable, prompting two anonymous and very generous donors from the Marina district to give a large sum towards its purchase. City officials eyed a fireboat retired from Vancouver, British Columbia, Canada. The vessel had been declared surplus and awaited dismantling in a scrapyard. With little time to waste, San Franciscans raised funds and fire commissioner Sharon Bretz

The *Phoenix* in its final configuration following rebuild in 1985. Changes included a redesigned after-deckhouse, mounting a 36,000-pound capacity crane. Wrap-around rub rails were added, and a remotely controlled elevated monitor replaced the tower. The 2,000-gpm monitor can be raised 48 feet above the waterline. Courtesy of Curt Elie Collection.

With black-painted deck and hull highlighted in blue and yellow during the 1970s, the *Phoenix* wears markings of the Port of San Francisco. Courtesy of Rich Schneider.

expedited the purchase in 1990. The next obstacle was getting the fireboat to San Francisco, a distance of 1,200 miles. Three pilots from the San Francisco Bar Pilots Association volunteered to crew the vessel, along with the very capable engineer of the *Phoenix*. The former Vancouver Fireboat 2, which had served the Canadian city since 1951, was made seaworthy, fueled, and loaded with supplies.

The long journey was anything but a pleasure cruise. With few comfortable furnishings aboard, the crew weathered a storm in 24-foot seas and endured hydraulic and electrical failures before being welcomed by the *Phoenix* and the people of San Francisco.

Sidestepping the limelight, the fireboat's benefactors retained anonymity, requesting instead that a contest be held for San Francisco school children in naming the boat. Six-year-old Christopher Smith's entry, *Guardian,* was selected as the winner.

The former Canadian fireboat *Guardian* has a stout appearance, which is backed by its ability to pump 24,000 gpm. The 12-foot catamaran-type skiff on the afterdeck stands ready to transport hose to supply shore-based units of the Portable Water Supply System. Courtesy of Rich Schneider.

The *Guardian* was built in 1951 by Canada's Yarrows, Ltd. Displacing 188 tons and measuring 88 feet long, it features four monitors, one of which is atop an enclosed tower, and a twenty-four-outlet discharge manifold. Original powerplants were five Kermath 12-cylinder 525 hp gasoline engines, two for propulsion and three to drive its pumps, which in 1972 were replaced by five GM Detroit diesel 12-cylinder 535 hp engines. The switch increased its gpm rating from 18,000 gpm to an impressive 24,000 gpm.

Both the *Phoenix* and the *Guardian* are berthed at Station 35, where ten assigned personnel staff an engine, plus the two fireboats.

Relying on a federal grant that months earlier had been delayed, San Francisco officials, as of this writing, await delivery of the city's first new fireboat in more than half a century. Requests for proposals went out in mid-2011, followed in 2012 by design of the fireboat from Jensen Maritime Consultants. The contract for the vessel was signed with Vigor Fab of Seattle in July 2014, with delivery slated for late 2015. Weighing 300 gross tons, the 88-foot fireboat is to be capable of pumping 18,000 gpm, with power provided by a pair of 793 bhp Caterpillar C-18 engines. In "super-pumper" mode, it will pump 12,000 gpm through numerous manifolds. Its crew of seven will include four firefighters.

Despite modernization, San Franciscans rely on protective systems developed from lessons learned during the early twentieth century. And the shining constant throughout the great city's history has been the courage and dedication of its firefighters.

Mare Island

Twenty-five miles northeast of San Francisco lies Mare Island, where the first US Naval base on the Pacific Ocean was established. The fire department of Mare Island Naval Shipyard at Vallejo was the first fully paid department in California, its establishment in 1854 having coincided with the onset of shipbuilding on the island. Although this civilian department was trained to fight both conventional and ship fires, it functioned separately from the US Navy's Mare Island Fire Fighter's School. The first recorded use of the department's waterborne firefighting capability is during WWII. Assistance in fighting major fires came

One could only wonder as to the stability of this arrangement when the elevated monitor delivered a high-pressure stream. This motor launch was modified to serve as the first fireboat of the fire department of Mare Island. Courtesy of US Navy.

from the base Marine barracks, US Coast Guard boats, US Navy tugs, and mutual aid from the Vallejo Fire Department. In 1943, the department acquired a 40-foot motor launch powered by a 105 hp diesel engine. Two 500-gpm centrifugal pumps supplied a portable monitor and a unique bow mast monitor incorporating a perch for the operator.

During the postwar period, the Mare Island Fire Department took possession of two surplus 40-foot Wheeler boats, which were supported by the navy 260-ton harbor tug USS *Dekaury* YTB-178 beginning in September 1950. The *Dekaury,* which was later re-designated YTM-178, was built in 1942 by Gibbs Gas Engine Co. of Jacksonville, Florida. Although the powerful tug's home port was San Francisco, it was moored at Alameda. When struck off charge from the navy in September 1985, it was sold to become the fireboat of Stockton, California. The Wheeler boats were removed from service during the mid-1960s, with waterborne fire protection of Mare Island assumed by navy and Coast Guard tugs, along with firefighting-equipped vessels operated by petroleum sites.

Since the Mare Island Fire Department planned to maintain a waterborne firefighting capability, officials looked at waterjet-powered boats, which were coming into vogue. The 21-footer selected for the department not only had tremendous punch at 3,600 gpm, it came with an interesting history. In 1955 Endura Craft began building fiberglass boats as a division of Dorsett Plastics of Santa Clara, California. They were called Dorsett boats beginning in 1958. Meanwhile, at San Rafael, California, in 1947 Victor and Ray Lenci, along with Ralph Cain, started Trailorboat Engineering Company. The trio began with an aluminum boat that incorporated a detachable wheel on the keel; hence the name Trailorboat. During the late 1950s, they began building a wide range of fiberglass boats, including fireboats. In 1965 they sold the fiberglass business to Dorsett, and in 1977 sold the aluminum business, which became Klamath.

The Mare Island Fire Department took delivery of its Dorsett fireboat in January 1966, which on February 16 was christened *James D. Greig,* after the former chief of the department. Satisfaction with the boat ended in 1982 when the navy converted the vessel from waterjet to diesel power. During trials with the new powerplant, the boat returned to shore full of so much water that firefighters felt the navy must have intended the boat for use with its submarines. The boat made its last trip down the Napa River in 1983 and never returned to the water. The *Greig* was the last fireboat operated by the Mare Island Fire Department, which was disbanded when the shipyard closed in 1996.

Mare Island Fire Department's ex-US Coast Guard Wheeler boat tests it pumps and monitors in 1963. The fire tug in the background, believed to be the USS *Dekaury*, tests under-pier nozzles. Courtesy of US Navy.

The navy tug USS *Dekaury* (YTM-178), along with other navy and Mare Island Fire Department units, conduct hose relay training in January 1964. Courtesy of US Navy.

The unique, but short-lived Trailorboat had its origins in California during the late 1940s. Initially waterjet powered, the Trailorboat of the Mare Island Fire Department met with failure when converted to diesel power. Author's Collection.

Monterey

Monterey is on the southern edge of Monterey Bay on central California's coast. It has operated fireboats, mainly to protect its fishing industry, since 1948. That year, the Monterey Fire Department purchased a Trim Tug in San Francisco. After renovation, which included a new engine and installation of a 1,000-gpm fire pump, the boat was named *W. E. Parker*, in honor of Monterey's first fire chief, William E. Parker. The 45-foot vessel, with narrow 11½-foot beam, and powered by a 225 hp engine, was manned full time by two firefighters. Since the *Parker* often did not live up to expectations, it was replaced in 1962 by a Valco waterjet boat. Jet boats produced by Valco Boats of Fresno, California, were similar to those made by Buehler Turbocraft. The then-trendy craft used a 215 hp Interceptor engine, with the fire pump mounted forward of the engine, and the propulsion waterjet pump located aft. Although constructed of aluminum, seawater took its toll, and its waterjets often became clogged along famous Cannery Row. Costs and time spent on maintenance proved prohibitive and the boat was retired during the late 1970s.

Monterey's fireboat program was revived in 1998 with the purchase of an experimental fireboat built onto a rigid-hull inflatable Zodiac, which featured a bow monitor. The two-person boat served until 2005.

In early 2011, the Monterey Fire Department, with a Homeland Security grant, purchased a MetalCraft Marine Firestorm 30 rated at 1,500 gpm. Designated apparatus number 6451 and named *The Pearl,* the fireboat features twin 250 hp Yamaha outboards. Berthed at Coast Guard Pier, *The Pearl* is cross-manned by Engine Company 6412 in New Monterey.

In July 1953, crewmen aboard Monterey's fireboat *W. E. Parker* secure a burning vessel in the harbor. The *Parker*'s narrow beam is evident in this view. Courtesy of Monterey Fire Department.

In December 1962 the Monterey Fire Department purchased a new Valco jetboat to replace the *Parker*, which often did not live up to expectations. Although constructed of aluminum, sea water took its toll on the Valco, and its jets often clogged along famous Cannery Row. It was retired in late 1970. Courtesy of Monterey Fire Department.

Members of the Monterey Fire Department and Monterey Boat Works in 1948 pull the engine of their first fireboat, a retired Trim tug. Pumps were also installed. The boat was purchased to protect the city's fishing industry. Courtesy of Monterey Fire Department.

Eureka

Along the rugged Northern California coast, on the shores of Humboldt Bay, Eureka hails as the largest coastal city between San Francisco and Portland, Oregon. The bay is linked with the Pacific and maritime industries, mainly fishing. In July 1958, the Eureka Fire Department acquired through military surplus from Mare Island Naval Shipyard a 40-foot Wheeler Boat. After refurbishment by department members, which included the addition of a pilothouse, the fireboat was placed in service as the *Captain Pudgy Davis* on October 2, 1958. The 14-foot beam of the former Coast Guard boat provided ample space for five Chrysler Marine engines, four of which drove corresponding pumps that delivered a total of 2,400 gpm.

John McFarland's thirty-two years serving Eureka Fire easily qualifies him to characterize the *Pudgy Davis*, transcending the usual technical profile.

> When Ray Lay became Eureka's fire chief, one of his first assignments was to purchase a new apparatus fleet, including a new fireboat. The boat was refurbished at the Zerlang Shipyard at Fairhaven, just across Humboldt Bay. A highly skilled blacksmith named Jack Tank, who worked at the city's Central Garage, manufactured all of the hardware to outfit the vessel. He transferred to the fire department to be the Fire Alarm Technician and also became the primary boat pilot as a result.

Ocean birds seem unfazed by the awesome power of Monterey's Firestorm 30 fireboat. The MetalCraft Marine fireboat was purchased for the fire department in 2011. Prior to beginning service, it was christened *The Pearl*. Courtesy of MetalCraft Marine.

Eureka, California, fire department members greet their new fireboat in 1958. The 40-foot Coast Guard Wheeler boat was acquired from Mare Island Naval Shipyard. Courtesy of Humboldt Bay Fire.

The vessel had four 600-gpm pumps feeding a large manifold that supplied a 2,000-gpm monitor and six 2 ½-inch outlets. Tank added a wheel house. It resided at the Eureka Small Boat basin. These finger docks were home to the local fishing fleet. During storm warnings, all boats came into port to wait out the storm, and it was common to find extra boats tied up four deep on the fingers. Everyone respected the *Pudgy Davis* and never blocked its egress.

On one occasion, while fighting a warehouse fire at the foot of "C" Street, the monitor was rotated too far and blew all of the windows out of the wheel house. When the turret was quickly elevated, the 2,000-gpm stream hit the aerial platform operating over a block away. Due to the toll that quantity of saltwater took on the electronics of the platform, it was never fully restored.

There is more than one explanation of how the fireboat got its name. McFarland explained:

George "Pudgy" Davis was one of the volunteer fire officers of the turn-of-the-century, 300-man, Eureka Volunteer Fire Department established in 1864. As the story goes, when Eureka purchased its first motorized fire apparatus, a 1915 Webb chemical wagon, the city council determined it was simply too complicated to be operated by a volunteer. So Pudgy Davis was promoted to captain, assigned as operator of the new apparatus, and appointed as Eureka's first career firefighter.

Another version holds that big, strong, single Pudgy Davis was the heartthrob of every young lady in Humboldt Bay. The memory of Pudgy as a true gentleman and as the most eligible bachelor seemed sufficient to some for naming the fireboat after him. McFarland continued:

The red and white-trimmed boat displayed the seal of the City of Eureka and was moored at city slip no. 1, kept ready at all times hooked to shore power for onboard lights and heaters and was visited every day by the engine crew of Fire Station 4 for maintenance and training. The *Captain Pudgy Davis* proudly protected Humboldt for many years fighting many waterfront fires involving docks, warehouses and other vessels. The *Davis* usually first darkened the raging fire with the monitor and then tied up to the burning dock and pumped to 3-inch lines to supply sea water to the engines on the dock. The big issue I recall with pumping at capacity was the constant rotational shutdown of one pump at a time to clean out the sea chest on the intakes which quickly plugged with eel grass from the bay.

The *Captain Pudgy Davis* served until 1985, when it was sold to a salvage company. In 1990, the fire department purchased from Wilmington, Delaware, a used 2,000-gpm fireboat it named *Marine One*.

In 2011 the Eureka Fire Department and Humboldt No. 1 Fire Protection District merged to form Humboldt Bay Fire Authority. Since Eureka's Humboldt Bay Harbor and Conservation District assumed re-

Although still recognizable as a former US Coast Guard fireboat, Eureka Fire Department personnel added a pilothouse for better visibility and control. Fore and aft monitors were retained. Surrounding the *Capt. Pudgy Davis* were intricately woven rope fenders. Courtesy of Humboldt Bay Fire.

sponsibility for waterborne fire protection, acquiring the area's first new fireboat became a priority. In September 2012, Moose Boats of Petaluma, California, delivered a 44-foot catamaran fire-rescue boat to the district. Intended for use in firefighting, rescue, as a pilot boat in dangerous shallows, and as backup for Eureka's hydrant system, the vessel is stationed at Eureka's Woodley Island Marina. Twin 600 hp Cummins diesels allow speeds in excess of 40 mph; coupled to twin Hamilton jets with dual fire pumps, they deliver 3,000 gpm.

Through the years, the Sacramento Fire Department has operated a variety of watercraft alongside police patrol boats. Most have been rigid hull inflatable types, such as this pump-equipped Polaris pulled by a battalion chief's 1991 GMC. In the twenty-first century, four fire stations house boats, one of which is the all-aluminum *Ever Vigilant*, rated at 500 gpm. Author's Collection.

San Diego

As a large city featuring thirty miles of shoreline and a natural deep-water port, it is no surprise that San Diego's experience with fireboats goes back to the turn of the century, when the city operated a steam-powered fireboat. As often happens during historical research, presenting an accurate account of fireboats proves challenging since references mention not only different types of vessels, but conflicting dates of service. Such is the case with San Diego. However, we do know about the early days of San Diego fireboats through a 1957 interview with San Diego's fourth chief engineer Louis Almgren. A popular, innovative San Diegan who served as fire chief from 1909 to 1935, Almgren said:

> . . . In 1914 we established our fire department shop and we built the fireboat with the firemen at Station No. 6 at Columbia and Cedar. The firemen built that in their spare time. We put all the mechanics in the shop and they did all the repair work in the shop and built the fireboat there in a period of about six years. The fireboat was named Bill Kettner to honor our famous congressman. The fireboat was 65 x 18, the size of the hull. The pumps were designed and built right in the fire department. Walter Forward was the designer of pumps and the pumps were made and manufactured right in the shop—the pumps are still in service. That was the first gasoline-powered fireboat by fire departments in service at that time in the United States. It was commissioned June 30, 1919.
>
> It had ten streams and two deck nozzles. Later we used the fog nozzles about the year 1920. We had a demonstration given in Los Angeles by Chief Scott. He was fire chief of Signal Hill, Long Beach—got the idea of the fog nozzle, and we had a demonstration for all the chiefs of Southern California, and this was the beginning of the fog nozzle in 1920—the chief of the oil fields there thought of and invented the fog nozzle (Sprung, 2013).

San Diego's first fireboat was built in the fire department shop during the early part of the twentieth century. It was named *Bill Kettner* in honor of the San Diego resident and congressman, who was said to have been responsible for the US Navy presence in the region. Courtesy of Curt Elie Collection.

San Diego's claim to have operated the nation's first gasoline-powered fireboat has periodically been refuted, with credit instead going to the Jacksonville, Florida, Fire Department. Historical records substantiate San Diego's claim since Jacksonville's gasoline-powered fireboat began service in 1922; the Jacksonville vessel pre-dating San Diego's *Kettner* never left the drawing board. San Diego chose gasoline because coal was a rare wartime commodity on the West Coast. Gasoline also permitted use of a smaller crew, and less fuel consumption since steam-powered fireboats needed constant coaling to get under way. The *Kettner* used three 220 hp engines—an Atlas Imperial for propulsion and two Seagraves to operate three centrifugal, two-stage pumps at 4,500 gpm.

Since Chief Almgren had a nautical background, he joined shop superintendent Charles Langell in overseeing construction of the fireboat. The machine shop at Station 6 is also where Robert Ely developed the National Standard Thread adopted by fire departments across the US.

Eventually the *Bill Kettner* underwent extensive rebuild, with its imposing monitor tower removed and a steel superstructure added. Power was derived from three Atlas-Imperial diesel engines, two of which had the dual function of driving a pair of 115 hp Seagrave two-stage centrifugal pumps.

Reportedly, the San Diego Fire Department in August 1948 received a fire and police patrol boat named *Chief Almgren.* The twin-engine, pump-equipped boat was small, fast, shallow-draft, and carried high-pressure fog and carbon dioxide extinguishing equipment. With a crew of two, the boat was designed for protection of the marina, with duties including

After extensive rebuild, which included the conversion to diesel power, San Diego's *Bill Kettner* bore little resemblance to its original configuration. Besides removal of its imposing tower, the pilothouse was rebuilt, abaft of which is a recessed hose reel and foam proportioning tank. Long spray nozzles are attached to the side of the pilothouse. Courtesy of Curt Elie Collection.

San Diego's 65-foot fireboat *Louis Almgren* began service in 1961. Details about its origin are uncertain; however, its lines are familiar to those of large military yard tugs. The mast atop the afterdeck supports a boom for deploying the lifeboat and a monitor platform. A Mexican caricature decorates the funnel. Author's Collection.

patrolling the bay, yacht basin, docks, and piers.

To keep pace with the city's growth, the fire department decided to replace the 40-year old *Kettner.* A fireboat having all the characteristics of a large military yard tug was obtained, and on July 7, 1961, was christened *Louis Almgren,* in honor of the progressive chief; the 4,000-gpm vessel went in service on August 21. On September 20, 1965, the *Kettner* was turned over to the Harbor Port Authority, which assumed fireboat operations the following year, much to the dismay of firefighters.

Additional fireboats made an appearance in San Diego waters during the 1970s with the 42-foot, 4,500-gpm *Harbor Island* in 1971, and three pump-equipped 32-footers acquired throughout the decade. They were the *Point Zuniga,* the *Point Loma,* and the *Shelter Island.*

In the twenty-first century, the San Diego Fire-Rescue Department Lifeguard Services operates a variety of vessels, the largest of which are three 35-

Among San Diego Fire-Rescue's fleet of Harbor patrol boats are 35-footers equipped with 1,000-gpm fire pumps. Courtesy of Curt Elie Collection.

The fire department's all-red fireboats, the *Louis Almgren* and the *Bill Kettner*, stand out against the shoreline of San Diego Harbor in 1961. One can only guess why a vehicle is secured on the *Almgren*'s deck. Author's Collection.

foot boats equipped with 1,000-gpm pumps to supply a bow monitor or hose outlets. In addition, the San Diego Harbor Police operates patrol boats that do double duty as fireboats staffed by crews cross-trained as firefighters.

This pair of MetalCraft Marine boats of San Diego's Harbor Police perform double duty as fully equipped fireboats, each staffed by firefighter-trained crews and equipped with three fire monitors. The department's dive team operates two boats. Courtesy of MetalCraft Marine.

The fireboat *Almgren* in the painting by L. M. Bucher commemorating the 100-year anniversary of the San Diego Fire Department. Courtesy of Curt Elie Collection.

CHAPTER 8: PACIFIC NORTHWEST

Portland

Portland, Oregon, like most cities across the nation during the late nineteenth century, seemed to have been built to burn. The Great Fire of Portland, which broke out on July 4, 1866, destroying 1,800 buildings and leaving 10,000 residents homeless, was the greatest fire yet seen in a US city. From the ashes arose an alarm system and a water system, while volunteer firefighters, called Red Shirts, did their best to cope with the looming potential for conflagration. And Portland city officials took up the struggle of other American cities—organizing, staffing, and equipping a paid fire department. Although that became reality in 1883, two decades would pass before plans were made to fight large fires from the water. Those who knew better were bewildered because running through the center of the city is the Willamette River, which flows into the mighty Columbia River, dividing Oregon from Washington State. Not only did downtown Portland grow up fronting the banks of the Willamette, miles of riverfront were lined shoulder to shoulder with structures made of fir, the construction material of choice since it was plentiful in the region; and the maze of wooden docks was soaked with petroleum product, coal tar creosote, and pitch.

Portland's first fireboat resulted from the vision of chief engineer David Campbell, a progressive leader who understood the need to place technology in the hands of the fire department he had molded. Construction of the 105-foot, wooden-hulled vessel began in 1903 at Portland's Willamette Iron & Steel Company. Relying on oil to fire her steam engines, the boat could pump 6,200 gpm to supply three fixed monitors, one of which was atop an elevated standpipe.

Portland, Oregon's, first fireboat, the *George H. Williams*, in 1905, the year after it began service. Overall white livery was common on West Coast fireboats during the first part of the twentieth century. Smoke billowing from her its funnels signal that oil-fed boilers are providing maximum power to pump through the three monitors, one of which is an elevating standpipe. Courtesy of Rick Howard Collection.

Contrary to claims by the New York Fire Department in 1908, Portland was the first American city with a fireboat that featured a stationary steel monitor tower. At its launching on February 27, 1904, the vessel was christened *George H. Williams*, honoring the Oregon mover and shaker, who had held the positions of Portland mayor, senator, Supreme Court justice, and attorney general. The *Williams* began service in May 1904, responding to twenty-nine alarms by year's end.

Within a few years, plans were under way for a second fireboat, which was designed by Chicago naval architect Theodor Knudson to be 125 feet long, with a 27-foot beam. The 242-gross ton vessel was steam driven by twin compound engines supplied by twin oil-burning Ballin boilers producing 1,200 horsepower. It was assembled in Ferrysburg, Michigan, in 1912 by Johnston Brothers Company of riveted iron-nickel plates on steel frames produced in Illinois. The vessel was knocked down, crated, and shipped to Portland to be reassembled at the yard of Joseph Supple by the Smith & Watson Iron Works, which produced many of the city's cast iron storefronts and hydrants. The vessel bristled with monitors supplied by pumps that gave a total output of 9,200 gpm. It underwent testing in May 1913 before being placed in service, and in December it was christened the *David Campbell.* Both boats had rough beginnings. In September 1904, when the *Williams* closed the distance with its first waterfront blaze, the stoker failed to light a boiler and water only dribbled from its monitors when they were aimed at the fire. When the boiler was lit, the powerful stream pushed the boat into the middle of the river, causing another delay while a great deal of real estate went up in smoke. As expected, the media had a field day. The *Campbell* came in for its own share of trouble. On July 29 of its first year, with monitors working over the burning Steel Bridge, it crashed into the bridge, damaging the forward monitor. Its bad luck didn't end there. On August 4, its crew was unable to raise a vacuum to allow pumps to put water onto a blaze that was later put out by land companies. One can only imagine the ribbing that fireboat crews endured from their land-based counterparts. With the new fireboat protecting the riverfront, the *Williams* underwent overhaul in 1914, emerging with a single, more powerful steam engine and increased pumping capability. Both fireboats gave good service, often backed up by Harbor Patrol boats that had limited pumping capability.

Chief Campbell was killed in a fire on June 26, 1911, but his advocacy of gasoline power remained strong. Not only had he brought the department into the motorized era in 1909, he embraced technology, making every effort to apply it to the fire service. As the waterfront grew with the city, the fireboats *Williams* and *Campbell* plied Portland's rivers. During the mid-1920s, Chief B. F. Dowell, who had been chief

The *Williams* in 1915 after it was fit with a single, more powerful steam engine during overhaul. Two lifeboats were added to the deck afterhouse, along with a boom for their deployment. Courtesy of Rick Howard Collection.

Launched in 1913, Portland's second fireboat was the *David Campbell*, rated at more than 9,000 gpm. Ten of its monitors are visible. The *Campbell's* keel and rub rail were brown and her funnels light gray. Courtesy of Rick Howard Collection.

since Campbell's death, issued specifications for three new fireboats to replace the pair of steam-powered boats.

> Each fireboat will be 85 feet long overall and will have a draft of only five feet, six inches so it can get close to a fire on the waterfront. The boats will be low, 12 feet, 6 inches, so they can pass rapidly under bridges without waiting for draws to open. They will have speeds of 18 to 20 mph and will be capable of pumping 8,000 gallons a minute each, at 200 pounds pressure. All controls will be in the pilothouse, so that one man can operate the boat and its pump at the same time.

His successor, chief engineer Leo G. Holden, made it official with this statement in the Portland Bureau of Fire 1926 annual report:

> At this time I wish to recommend that the two fireboats be replaced with three gasoline motor driven boats, each with a capacity of 6,800 gallons per minute. I have plans prepared for boats of this type.

Supporting his proposal for gasoline versus oil, Holden added:

> In 1925 the two boats consumed fuel oil as follows: 318 barrels were used for the purpose of propelling the boats to oil docks for fuel oil; 7,552 barrels were used in fighting fires and also for keeping up steam while at docks and while going to and fighting fires.

Naval architect A. D. Merrill of Portland designed the vessels to meet Bureau of Fire's specifications, laying out a less-than-battleship design to be operated by the smallest number of crewmen possible. Portland's Baker Construction Company got the contract for building the trio. Displacing 120 tons, each boat had beneath their decks two 565 hp, eight-cylinder Sterling Viking II gasoline engines. The engines also drove the pumps, in addition to a pair of 425 hp six-cylinder Sterling Vikings directly connected to two Byron Jackson 2,180-gpm pumps and two 2,680-gpm pumps. Each boat could pump approximately 10,000 gpm at 200 pounds.

As advertised, Portland's new triplets featured complete, one-man control from the pilothouse. Besides the pilot, the small crew consisted of two men in the engine room and two on deck. The first of the steel-hulled trio was named the *David Campbell (II)* and began service at the end of 1927 as Fireboat 2. Next was the *Mike Loudenklos,* which began service on January 7, 1928, as Fireboat 1, followed by the *Karl Gunster*, in service as Fireboat 3 on February 2. All had been named after Portland firefighters killed in the line of duty.

Portland's Bureau of Fire in 1961 initiated a costly modernization program that included updating the fireboats to maximize their ability to protect thirty miles of harbor frontage. Nearly all of the boats' operating systems required overhaul. Eight 550 hp Hall-Scott V-12 gasoline engines were purchased from military surplus to replace the Sterling Vikings. Also installed in each vessel were two 3,500-gpm De Laval pumps driven by pumping engines, supplemented by a pair of 2,750-gpm pumps driven by the propulsion engines when the boat was not under way. The four pumps boosted each boat's pump output from 10,000 gpm to 12,500 gpm. Steering was converted to pneumatic power, as was the midship monitor,

When raised, the single-section aerial ladder installed on Portland's three fireboats in 1961 positioned an 850-gpm nozzle 43 feet above the waterline. The aluminum pipeway supplying the nozzle is visible in the bed of the ladder of Fireboat 1, the *Mike Laudenklos*. Courtesy of Rich Schneider.

The *Mike Laudenklos* was one of three 10,000-gpm fireboats built for Portland's Bureau of Fire during the 1920s. The trio featured a massive monitor amidships. Courtesy of Rick Howard Collection.

The *David Campbell* was the first of Portland's new trio of fireboats delivered during the late 1920s, and the second fireboat named in honor of the department's progressive chief. The *Campbell* was designated Fireboat 2, while the *Karl Gunster*, pictured here during the 1950s, was designated Fireboat 3. Courtesy of Rick Howard Collection.

All three fireboats had two banks of six gated 3½-inch hose outlets. Pictured here, with ladders secured to the railing, is the starboard manifold of the *Laudenklos*. Courtesy of Rick Howard Collection.

The *Karl Gunster* displays the change in markings after Portland's Bureau of Fire assumed harbor patrol duty in 1973. Also noteworthy is the stretch of the tower added in 1961, along with a red and white scheme. Courtesy of Rick Howard Collection.

Portland's second fireboat named the *David Campbell* following its 1975 overhaul. Not only was it repowered with diesel, boosting output to 14,000 gpm, the pilothouse was rebuilt. The rose added to the pilothouse represents the city's nickname Rose City. The powerful amidships monitor with 4-inch-diameter tip was a carryover from the original fireboat *Campbell*. Courtesy of Rick Howard Collection.

allowing one crewman to operate the massive appliance, which previously required two or three crewmen. A significant addition to each boat was a steel 30-foot, electric-powered aerial ladder, with controls at its base and at the tip. The ladder was intended not only for access to ships and docks, but to connect hose to sprinkler systems of waterfront buildings, and to dry pipelines beneath the East Bank Freeway along the waterfront. Other improvements included more lighting plants, a new electric heating system, and more efficient engine room ventilation. New fireboat stations, including a floating platform moored at Terminal No. 4, complemented the three revamped fireboats.

Following the trend begun by fire departments across the nation during the 1960s, Portland in January 1972 purchased a waterjet-propelled fireboat. The 43-foot aluminum-hulled vessel was built by the Rohr Corporation. Rohr was a major aircraft concern that embarked on high-speed marine technology and shipbuilding during the late 1960s. The fireboat began service on July 1 as Fireboat 4, although christened *Virgil Spencer* in honor of Lt. Virgil L. Spencer, who was killed at a fire in October 1966. Drawing less than 3 feet of water, the *Spencer* easily accessed places that were denied the big boats.

A 400 hp Detroit diesel engine powered twin Jacuzzi waterjets. Although built for speed, the

Spencer could deliver 6,000 gpm at 150 psi, thanks to two Jacuzzi turbine pumps, each powered by a separate 400 hp diesel engine. The pumps supplied two deck monitors, along with six discharge outlets.

Like its early predecessors, the *Spencer* had a rough start. During its public debut, a debris-damaged pump housing caused the hull to fill with water, swamping the vessel by the time its crew maneuvered it back to the dock.

In 1972, one of the *Spencer's* jet pump casings cracked from debris picked up in the river. The crew got the boat back to the dock and tied off the bow before it sank. The boat was raised, repaired, and served many more years. Courtesy of Rick Howard Collection.

In a contentious move that harbored strong political implications, the Bureau of Fire in 1973 assumed the duties of the Police Harbor Patrol. The change meant that two harbor patrol boats would be transferred to the fireboat fleet. They were the 56-foot *Capt. Karl Prehn*, built in 1957 by Albina Engine & Machine, and the 33-foot *L.V. Jenkins*, built in 1968 by Pacific Tank & Construction Co. Both were steel-hulled, diesel-powered vessels, with only the *Prehn* equipped with a 500-gpm pump and deck monitor. In 1976, both were given significant firefighting capability with the installation of a 1,000-gpm pump, deck monitors, and discharge outlets.

Included in the move was an 18-foot fiberglass-hulled patrol boat, leased to the police department from its builder, Bramco. Named *Sea Swirl,* the boat became Fireboat 7, while the *Prehn* became Fireboat 5, and the *Jenkins* Fireboat 6.

The *David Campbell* in 1975 was rebuilt with two 500 hp Cummins diesel engines for propulsion and for driving two 3,500-gpm De Laval pumps. An identical pair of diesels was directly connected to two more 3,500-gpm pumps, boosting the *Campbell's* output to 14,000 gpm.

As the largest and most powerful of the boats inherited from the police department, the *Prehn* remained in service until 1983. Its replacement, the second fireboat to be named *George H. Williams*, arrived from Work Boats Northwest shortly thereafter, beginning service as Fireboat 2 on October 20. With the decline in large fires and a greater need to access waterfront property, the *Williams* reflected the preference for smaller, faster, and more maneuverable fireboats. Weighing 10 gross tons, the 40-foot *Williams* sported an aluminum hull, with power derived from a pair of Detroit 410 hp diesel engines, each driving a 1,250-gpm Waterous two-stage centrifugal pump.

Less recognized, although equally important among Oregon's maritime response resources, was a 27-foot Monark 1,000-gpm fireboat built for the US Army Corps of Engineers after the 1980 eruption of Mt. St. Helens. First used at Spirit Lake, the diesel-powered, waterjet-driven vessel suffered hull damage and was run aground. It was later given to Portland's Bureau of Fire under the Federal Excess Property Program (FEPP). The boat was repaired and assigned to St. Helens Fire Station on the Columbia River, where it served fifteen years. When Portland received a grant to replace the boat, FEPP offered it to the Lakeside, Oregon, fire department, which took delivery in September 2008.

A partner to the fireboat *Williams* arrived in 1996 from Rozema Boat Works at Mt. Vernon, Washington. The aluminum 42-footer featured twin 550 hp Detroit diesels, each driving an Aurora 2,050-gpm pump.

Filling out Portland's fleet was a 30-foot, 400-gpm fast response boat built by Aluminum Marine Construction for Portland International Airport. Additional small rescue boats were dispersed throughout Portland's waterways.

Like many major cities, Portland used land apparatus in conjunction with fireboats. The Portland Bureau of Fire acquired this 1947 American LaFrance through annexation of the City of Multnomah in 1951. The truck was converted by the department shop into a manifold unit. Courtesy of Chuck Madderom.

Portland's Manifold 3 was a 1968 GMC/Wentworth & Irwin with pump panel and booster reel. Courtesy of Chuck Madderom.

It took a tragedy to expose a weakness in Portland's river rescue response and bring about the purchase of a fast, powerful boat in 2010. For the first time in Portland history, a fire department vessel was not named after a former firefighter. Instead the boat was named *Eldon Trinity* after two young children who were thrown off the Willamette River Bridge by their mother intent on killing them. Trinity was rescued and survived, but her younger brother, Eldon, drowned. Since it took the powerful, but slow-moving *David Campbell* forty-four minutes to arrive, the new boat was stationed at a central city fire station for quick responses.

A name change from Portland Bureau of Fire to Portland Fire Rescue in 1988 not only reflected its wide range of services, it signaled a significant rise in maritime emergency responses, which, as of this writing, surpass 300 annually. To meet these challenges, plans were made to purchase two high-tech fast-response fireboats. A contract was let to Oregon Iron Works of Clackamas for the 55-footers powered by twin 1,085 hp engines driving Rolls Royce waterjets. Designed by Jensen Maritime Consultants of Seattle, the aluminum 8,000-gpm boats feature a 1,500-gpm bow monitor and a 4,000-gpm monitor atop the cabin. Noteworthy features include hydraulic bow thrusters and an automated compensation system for recoil from monitors. Speeds of 45 mph permit quick responses along Portland's long stretches of waterways. Since it opened its doors in 1984, Oregon Iron Works has built hundreds of high-performance military boats, with this pair being their first fireboats.

In July 2015, the pair was christened with chinuk wawa names honoring the original river people of Portland's waterways; *kwansum,* meaning "always ready, always there," and *skukum ats,* meaning "strong sister."

The *Karl Gunster* and the *Mike Laudenklos* were sold in 1977; the *Gunster* still lies derelict on land at a Yaquina River yard. The second *David Campbell* still serves in reserve, while the original *Campbell* went on to a colorful and storied career. Retaining only its official number 211065 and Johnston Brothers hull number 52, it was sold to Foss Towing in 1928. The following year it was purchased by Kanaga Ranching Company, converted from steam power to sail-diesel power, and renamed *Chief.* Revamped as a freighter and renamed *Aleutian Native*, it hauled furs and cargo between Kanaga Islands in the Aleutians and Seattle. On its maiden voyage in stormy seas, a 100-foot wave swept clean its deckload of lumber. In 1932 it went to Petroleum Navigation Company, which converted it to an oil tanker and renamed it *Express* in 1939. After supplying fuel and cargo to coastal Alaska during WWII, it was modified for towing barges to supply fish canneries; between tows, it served as a Bering Sea crabber. Purchased in 1956 by the Upper Columbia River Towing Company for towing log rafts, the *Express* returned to crabbing and salmon tendering during the 1960s, providing this service until its eighty-ninth birthday in 2001.

The fireboat *David Campbell* in 2007 renders a patriotic salute to the guided missile frigate USS *Vandegrift* (FFG 48) as it passes Willamette River Bridge during the 100th Annual Rose Celebration. Courtesy of US Navy.

Portland's twin high-tech fireboats after being christened *kwansum* and *skukum ats* in 2015. Author's Collection.

Seattle

Since its earliest days, Seattle's waterfront has been a major factor in its economy. The Port of Seattle has been one of the key harbors on the West Coast since the late 1800s, having long thrived as a major lumber port. And like many cities across the nation during that period, Seattle suffered a great fire. The Great Seattle Fire broke out on June 6, 1889. Before its destructive sweep abated, twenty-nine square blocks, including nearly all the city's waterfront and business district, had been destroyed. Amazingly, no lives were lost.

Then-mayor and shipbuilder Robert Moran implored distraught citizens that rebuilding the city should include a fire department to safeguard new buildings. The master plan to create a professional, full-time fire department included the design and construction of a fireboat to protect the waterfront that was certain to rise again. Within one year, Seattle had five new firehouses, a box alarm system, and the keel was laid for the first purpose-built fireboat on the West Coast.

The fireboat was designed by renowned naval architect William C. Cowles, and the job of building it went to Seattle Shipbuilding & Dry Dock Co. Mayor Moran had a vested interest in the company, and when his two-year term ended in 1890, he returned to shipbuilding. In 1906, Seattle Shipbuilding & Dry Dock became Moran Brothers Shipyard. Aboard the vessel for its celebrated trial run was T. J. Conway of the Pacific Insurance Association, who said, "She did very well, splendidly, in fact, and I shall feel justified in recommending a liberal reduction in insurance rates here."

The wooden-hulled, coal-fired, steam-powered fireboat was placed in service as Engine 5 at the foot of Madison Street on January 3, 1891. Eventually, it was named *Snoqualmie* after a Pacific Northwest Indian tribe. Until its arrival, the Seattle Harbor Patrol, which had been in existence since 1877, operated vessels that were marginally equipped for fire suppression.

Lengthwise, *Snoqualmie* measured 90 feet, with a beam of 22½ feet and drew 10 feet of water. It boasted 7,000 gpm, thanks to two Clapp & Jones pumps.

The rapid growth of Seattle and its waterfront justified the acquisition of a second fireboat. Anticipating long-term growth, officials decided to pull out all the stops and aim for a battleship-class boat on a par with *The New Yorker* and San Francisco's twins, the *Scannel* and the *Sullivan.*

As the vessel took shape on the drawing board of naval architect Eugene L. McAllaster, it evolved into more of a battleship than anyone had imagined. It was not lost on the ship's designers that Seattle was called Lumber Supplier to the Nation. Although

Seattle's first fireboat, the *Snoqualmie*, was built after the June 6, 1889, Great Seattle Fire. The wood-hulled Engine No. 5 was 90 feet long and pumped 5,765 gpm. It went into service on January 3, 1891. Seattle's first on-duty firefighter fatality was a crewman. The *Snoqualmie*, rebuilt in 1090 and re-designated Engine 31, served Seattle for forty-four years until retired on September 25, 1935, a victim of the Great Depression. Courtesy of Museum of History & Industry.

only twelve miles long, the city had seventy-five miles of waterfront, most of it occupied by the lumber industry along Puget Sound. Since wood was plentiful and inexpensive, it was the material of choice for building sawmills, drying kilns, and warehouses. The sound's topography was formed by mountains of lumber. Even the ships used for commerce at the sound were made of wood.

Combatting such gargantuan, fuel-loaded blazes required immense volumes of water at great pressures. A fireboat capable of delivering such knockout power had to be able to withstand tremendous punishment for long periods of time. The stout vessel would be of all-steel riveted construction. To operate in Seattle's shallow waters, McAllaster gave it a relatively flat bottom without a distinct keel. His massive, reverse-angle bow was an even more radical deviation from conventional ship design. Made of reinforced iron, the bow's sole function was to act as a cleaver when ramming and sinking a burning lumber schooner that resisted extinguishment and threatened the waterfront. No record exists of another fireboat built with a ramming bow.

The fireboat would stretch 120 feet and measure 28 feet in the waist. Although weighing a hefty 322 gross tons, it would draw 9½ feet of water, bringing it within shallow draft limits. In powering the vessel, McAllaster passed on emerging turbine-centrifugal technology, opting instead for proven steam boilers and piston pumps. Four oil-fired Mosher water tube boilers drove twin screws and three American LaFrance duplex piston pumps, each rated at 3,000 gpm. The massive boat churned through Seattle's waters at a mere 12 mph, but speed was not as essential as its ability to battle fire.

For knockdown power, McAllaster added eight large monitors—three on each side, one aft on a raised platform, and the largest atop the pilothouse. That monitor alone, with its 7,000-gpm wallop, would become known for smashing through buildings, snapping timbers, and knocking over burning boxcars loaded with lumber.

On July 3, 1909, the mighty fireboat slid the ways at Richmond Beach Boatbuilding Co. of Washington. It was named *Duwamish* after the river that coursed through Seattle and the Native Indian tribe of western Washington. After outfitting, sea trials, and crew training, *Duwamish* in 1910 took up station at Engine 5, replacing the *Snoqualmie*, which went in for rebuild. The work included conversion of its steam engine to burn oil. It was outfitted with two Taylor water tube boilers and two Clapp & Jones duplex piston pumps rated at 3,000 gpm each. The *Snoqualmie* was then placed in service as Engine 31 at Lake Union.

The *Snoqualmie's* hose wagon, horse, and crew pose at Fire Station 5 built after the Great Fire of 1889 to service the new fireboat. This was the first of several stations built at the foot of Seattle's Madison Street. Courtesy of Anders Wilse.

Other than the sweeping sheer of its hull, the fireboat *Snoqualmie* bore little resemblance to its original form after rebuild in 1910. The pilothouse underwent significant redesign to accommodate the funnel of a second boiler. Much of the wrap-around bulwark was removed and the number of monitors was greatly increased. Courtesy of Puget Sound Maritime Historical Society.

Little separated battleship and potential foe. Adjacent to the berth of the *Duwamish* and Station 5, which housed Hose 5, was the largest wooden structure on the Pacific Coast: the Grand Trunk Pacific Dock—500 feet long, over 100 feet wide, supported by thousands of creosote-treated pilings, and incorporating a three-story wooden warehouse topped by a 108-foot tower. The imposing complex was built with four million board feet of lumber. Void of any fire protective measures, it was the proverbial disaster waiting to happen. Foreshadowing disaster, nearby

After conversion to the freighter *Robert Eugene* during the late 1930s, the *Snoqualmie* was no longer recognizable as a fireboat. More than thirty years after the conversion, it burned and sank at Kodiak, Alaska. Courtesy of Puget Sound Maritime Historical Society.

Seattle's fireboat *Duwamish* was new in 1910. Side-by-side funnels were an unusual arrangement, compared to in-line stacks such as those of the *Snoqualmie* in the right background. Author's Collection.

Hose 5 was called on daily to find and extinguish smoldering fires caused by smoking materials that fell into the joints of the wooden decking. That smoking was even permitted in such flammable surrounds is, in itself, unfathomable. Seattle firefighters had yet to experience their first general alarm fire, but they knew that when it came, it would be at the Grand Trunk.

On July 30, 1914, the inevitable happened. Workers at Grand Trunk spotted smoke near where the freighters *Athlon* and *Admiral Farragut* were taking on cargo. A worker alerted Hose 5, which went to the fire. Feeding on the massive fuel load, the fire quickly worsened. The crew of *Duwamish* headed it into the slip but was delayed by the *Athlon* and *Farragut* fleeing to open water. The mammoth structure became super-heated, reached ignition temperature, and flashed over. Seattle's first general alarm was sounded as intense heat radiated across the 170-foot-wide slip to ignite the Colman Dock and its warehouse. The steel *Duwamish*, caught between the twin infernos and thick smoke, became too hot to touch, despite the pounding streams from its huge monitors.

The *Snoqualmie* arrived, along with the US Revenue cutter *Unalga*, and began an attack in another slip to save a dock and warehouse. Two of Hose 5's three firefighters became trapped and severely burned but managed to get into the water, where they were rescued by fireboat crews. Their hose wagon had dropped through the burning dock into the water. Five persons, including Hose 5's driver, are known to have died. Officials were certain that were it not for the powerful streams of the *Duwamish* and *Snoqualmie*, losses would have been higher.

Countless blazes on the waterfront and aboard vessels were the stock and trade of the two fireboats. With plenty of work for a third battleship-class fireboat, a 124-foot riveted steel vessel was designed by Seattle's W. C. Nickum and built in 1927 by Pacific Coast Engineering of Oakland, California. Weighing 196 gross tons, it was powered by a 350 hp Winton six-cylinder gasoline engine driving triple screws. Another six identical engines drove the pumps, with two available for propulsion. Six Byron-Jackson Multiplex four-stage centrifugal pumps delivered a total of 12,000 gpm. A pair of Byron-Jackson 1,000-gpm centrifugal pumps was powered by two Fordson gasoline engines.

Named *Alki*, after the city's original settlement that was annexed to Seattle in 1907, the powerful fireboat began service on December 10, 1928, at

Smoke spewing from the 45-foot high funnels announce that *Duwamish's* oil-fired boilers are hard at work to power pumps delivering water to seven monitors. The pilothouse monitor, nearly 8 feet long with 6-inch-diameter tip, was known for its ability to throw a knockdown stream hundreds of feet. The unique projecting bow was designed to ram and sink burning lumber schooners if the fire could not be extinguished. Courtesy of Museum of History and Industry.

When the 124-foot long *Alki* was built in 1927, only four fireboats in the nation were longer. The tower, which was 42 feet above the waterline, was removed in 1994 after sixty-five years of service. Courtesy of Puget Sound Maritime Historical Society.

During rebuild of the *Duwamish* during the 1930s, her eight monitors were heightened to better attack large fires along Seattle's waterfront. Visible on the raised deckhouse is a rectangular pipeline supplying monitors. Courtesy of Museum of History and Industry.

Station 5, shifting the *Duwamish* to Engine 31 and the *Snoqualmie* to Engine 39. The *Alki* proved its worth the following year when the Union Pacific Dock burned. Its crew not only held the fire in check, they got close enough to the blazing pier to rescue four trapped firefighters.

During the 1930s the *Duwamish* underwent rework to replace large-diameter pipe loop. All eight monitors were heightened, and the aft one, along with the giant pilothouse monitor, were surrounded by steel platforms. A boom and hoist was added aft.

Throughout its long career, the *Snoqualmie* not only fought fires, it went to the rescue of ships in Puget Sound, sometimes performing salvage duty by pumping out sunken vessels. Weary after forty-five years of service, in 1935 it was placed in storage and sold the following year. After a major rebuild, which included the switch to diesel power, it became the *Robert Eugene.* After thirty years of hauling freight in Alaska, it caught fire and sank at Kodiak.

During mid-1948, *Alki's* powerplant was converted to direct drive diesel with twin screws using two General Motors 500 hp engines. Power for six pumps now came from twelve 196 hp diesels, with two driving each pump. New power for the salvage pumps came from two Westinghouse 20 hp electric engines. With an increased rating of 16,200 gpm, it was then the nation's most powerful fireboat.

Seattle's fireboat modernization program continued with the *Duwamish* in 1949. It was repowered with two 765 hp GE electric engines supplied from three generators powered by three 900 hp Cooper-Bessemer

Seattle's Harbor Patrol, which predates the fire department, included in its fleet vessels equipped for firefighting. Here, the *Duwamish* pumps in proximity to Harbor Patrol No. 7, an ex-army 65-foot T-boat. Author's Collection.

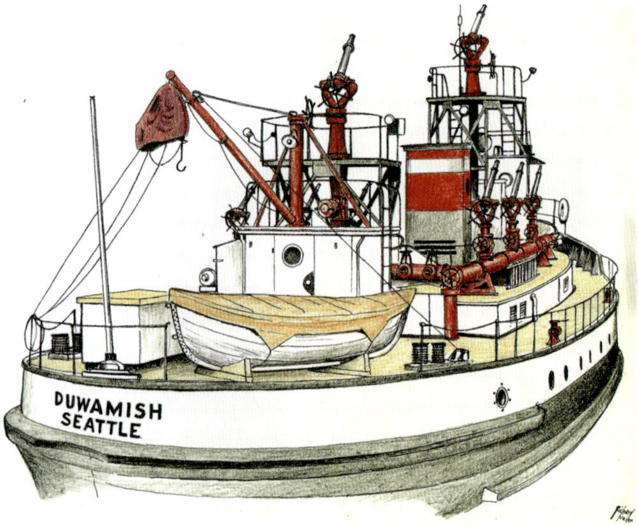

This rendering of the *Duwamish* by fire historian Richard Nailen not only clarifies the fireboat's details, such as the fire main loop, monitor arrangement, and boom, but the boat's color scheme. Courtesy of Dick Nailen.

Seattle was one of numerous cities whose fire departments made use of military surplus amphibious vehicles for fire and rescue duty. The Seattle Fire Department acquired this 1946 GMC DUKW-353 in 1965. It was painted red with yellow lettering and designated Apparatus no. 302. Courtesy of Last Resort Fire Department.

diesel engines. A pair of 11,400-gpm De Laval pumps was installed, powered by two 765 hp GE engines driven by two of the propulsion generators. With its new rating of 22,800 gpm, *Duwamish* claimed the title of most powerful fireboat in the world. Other significant changes were a redesigned pilothouse and removal of its ramming bow in favor of a conventional bow.

With three fireboats in service, Seattle Fire Department officials considered a smaller, more versatile craft for rescue work on the city's waterways. From military surplus they acquired a former US Coast Guard DUKW-353 amphibious vehicle, popularly called Duck. The 1946 GMC amphibian was the Coast Guard version built of aluminum to resist corrosion in its role of surf rescue. After taking delivery of the Duck in 1965, the fire department shop installed a 50-pound Purple-K dry chemical extinguishing system, along with a small fire pump and other firefighting and rescue equipment. The Duck began service as ARC-1 in 1966 as a special apparatus manned only during disasters and special events, such as hydroplane races. It remained on the inventory until May 1974, when it was sold.

Still in need of a small boat, the fire department in January 1976 acquired from the Seattle Harbor Patrol

This former police patrol boat was the Seattle Fire Department's first vessel, named *Leschi*. It served the fire department for five years before being returned to the police department. Courtesy of Rich Schneider.

a 30-foot Uniflite built in 1961. It was powered by a pair of Chrysler Marine 413 hp gasoline engines and equipped with a Hale 150-gpm portable pump. Named *Leschi,* the boat was assigned as Harbor 2 at Station 4.

In January 1982 the *Leschi* was returned to the Harbor Patrol. Since Seattle's waterways were used more than its underdeveloped roadways during the late nineteenth century, the Harbor Patrol predated the fire department. The patrol had acquired its first official vessel in 1908 and added boats during Prohibition. The Harbor Patrol remained a significant presence in Seattle waters, becoming an element of the Seattle Police Department during the 1960s. Throughout its history it has operated boats equipped with fire pumps in a cooperative effort with the fire department.

When the *Alki* went on reserve status in 1982 after more than half a century of service, plans were made for a new boat that would alternate reserve-active periods to extend the life of both vessels. During late 1984, the fire department retired the *Duwamish* as

Engine 4 and placed it in storage when it took delivery of a 92-foot fireboat built by Nichols Brothers Boat Builders of Freeland on Washington's Whidbey Island. With a wealth of experience in work boats, Nichols welcomed the opportunity to build its first fireboat, hull number S-70. Three 1,000 hp Detroit diesel engines powered triple screws as well as three 2,500-gpm Worthington pumps. Besides three deck monitors, the boat featured a Stang monitor atop the pilothouse that telescoped 50 feet above the waterline; two bow monitors were added for under-pier firefighting. Built into the stern of the boat was a ramp for launching a small rescue boat, a feature that would appear on fireboats of other large cities. When the fireboat began service as Engine 1 on January 8, 1985, proudly emblazoned on its tall, white-painted superstructure was the name *Chief Seattle.* That same day, the *Alki* was placed in reserve; three months earlier, the *Duwamish* had been placed in storage.

When rebuilt in 1948 with diesel direct-drive engines, Seattle's *Alki* was billed as the nation's most powerful fireboat. The stout tower was removed in 1994. Courtesy of Bill Hattersley.

The mighty *Duwamish* was a true firefighting battleship, designed to withstand all extremes while battling major blazes. The monitor mounted high atop the pilothouse alone delivered 7,000 gpm. It is seen here in its heyday at Seattle's Lake Union Pier. Author's Collection.

In its waning years, the *Duwamish* remained a formidable weapon against major blazes. Visible in this profile view is the extensive water main loop that supplied the monitors and discharge outlets. Courtesy of Fire Apparatus Journal Collection.

The pump engine telegraph in the pilothouse of the *Duwamish* was typical of those found in American fireboats. The pilot's positioning of the handle corresponded to an identical dial in the engine room, and rang a loud bell. The *Alki's* monitor tower is visible through the porthole. Author's Collection.

For nearly two decades, the *Chief Seattle* and the *Alki* protected Seattle's waterfront. The only change in marine operations was the addition of a Zodiac boat in 1998. The donated boat featured waterjet propulsion, which also supplied a prow monitor. After a large fire at Lake Union in 2002 destroyed thirty vessels, the *Alki* was permanently stationed on the lake at Fishermen's Terminal. The following year, a levy passed by voters allocating $167 million to upgrade fire department services not only helped finance modernization of the *Chief Seattle* and the purchase of new fireboats, it paved the way for *Alki's* retirement.

The first element of the upgrade package arrived from MetalCraft Marine during July 2006. Named *Fire One,* the fast attack fireboat began service as Engine 1 on March 6, 2007. The vessel would prove particularly valuable if both large fireboats were laid up. The 50-footer was powered by Hamilton waterjets driven by two 715 hp Caterpillar diesel engines. The same engines powered two Hale centrifugal pumps rated at 3,300 gpm each. *Fire One's* two primary monitors could draw from a 200-gallon tank of foam concentrate.

The wheelhouse, also called pilothouse, of Seattle's fireboat *Alki*. Flanking the wheel, or helm, are engine throttle pedestals and pump engine telegraph. Vital to steering and navigation, the nautical compass and navigation equipment is protected within the binnacle, immediately forward of the wheel, and in line with the helmsman's vision. Magnetic spheres on each side of the compass balanced magnetic disruptions caused by the compass. Author's Collection.

The Stang monitor atop the *Chief Seattle's* imposing superstructure telescoped 50 feet above the waterline. An equally powerful hand-controlled monitor is on ther open foredeck. Beneath the "wings" of the superstructure are mounted anchor flukes, a litter basket, and life raft. Courtesy of Bill Hattersley.

The *Chief Seattle's* sloped ramp for launching a small boat was adopted by other builders of large fireboats. Below the flying bridge on both sides are discharge manifolds for hose connections. Courtesy of Last Resort Fire Department.

The *Alki's* more conventional lines are obvious in this comparison view. The boom attached to the wheelhouse could be used for deploying a skiff or for salvage work. Compared to the *Chief Seattle*, the *Alki* uses a combination of fixed railings and bulwarks. Author's Collection.

Next in the lineup was a floating powerhouse that made the transition in 2007 from hull number 49 of Dakota Creek Industries at Washington's Port of Anacortes to Seattle fireboat named *Leschi.* Weighing 303 gross tons and measuring 108 feet long with a 27-foot beam, the Large Platform fireboat features four 1,555 hp Detroit MTU diesel engines below deck. Two drive the vessel to 16 mph, while two power four 5,000-gpm Fire Fighting System (FFS) centrifugal pumps. All monitors are from the Norwegian firm FFS, and all are remote controlled and capable of drawing from the 6,000-gallon foam supply. A 55-foot boom mounted aft incorporates a 70-foot extension ladder, along with a 1,500-gpm monitor. The vessel can be controlled from five positions—four in the pilothouse, which features all-around visibility, and outside at the afterdeck.

The *Leschi* takes its name from a chief of Washington's Nisqually tribe who was convicted of murder in 1858 and hanged. He was exonerated of the crime in 2004. The *Leschi* began service at Engine 4, eventually moving to Engine 5 on Elliot Bay, alongside the *Chief Seattle.*

Like its sister fireboat, *Chief Seattle*, abaft the *Leschi's* substantial superstructure is a sloped ramp for small boat operations. Visible here is the 55-foot boom, which incorporates a 70-foot extension ladder and a 1,500-gpm monitor. Courtesy of Last Resort Fire Department.

The overhaul and repowering of the *Chief Seattle* completed Seattle's fireboat fleet upgrade. Beginning in 2009, the firm of Guido Perla & Associates, comprising naval architects, and marine, mechanical, and electrical engineers, undertook management of the project, with the work accomplished at Vigor Marine of Everett. The overhaul, which began in January 2012, was expected to take the fireboat into the year 2033. The work included a new superstructure with a lower roofline and pilothouse with command center; complete replacement of electronic navigation and communication equipment, plus infrared surveillance gear; a new medical treatment room; and new rescue boat and boat recovery system.

The *Chief Seattle's* three diesel engines were replaced with pump drives having two 1,522 hp propulsion engines and new twin-screw drive. In addition, two 715 hp Caterpillar fire pump engines with four 2,500-gpm Hale pumps were installed, increasing total gpm output to 10,000 gpm. A new 900-gallon foam system was installed, along with Stang monitors, which included under-wharf monitors.

The new and improved *Chief Seattle* was delivered in February 2013 at Vigor's yard, and resumed duty as Engine 3. Three new, powerful fireboats allowed the fire department to rotate them through two-year maintenance cycles, while always having one operational. The trio also allowed the sale of the Zodiac boat and moving forward with the *Alki's* retirement. The city placed it on eBay auction in spring 2013, but bidders proved not to be viable options. It was sold in March 2013.

The *Chief Seattle's* pilothouse was a far cry from those of older fireboats it had replaced. Gone are the numerous pedestals with throttle and telegraph controls, and ornate helms. Courtesy of Last Resort Fire Department.

The *Chief Seattle's* 2012–2013 rebuild is expected to extend the vessel's life twenty years. The most recognizable change is a redesigned superstructure with new pilothouse. Courtesy of Rich Schneider.

The Seattle Fire Department took delivery of this MetalCraft Marine Custom 50 in July 2006. Capable of pumping 6,600 gpm, the boat is expected to provide waterfront coverage if the department's two large boats are in dry dock. Named *Fire One*, the waterjet-powered boat can reach a speed of 34 mph. Courtesy of Rich Schneider.

With a beam of 27 feet, the *Leschi* was called a Large Platform fireboat. It is rated at 20,000 gpm, with the bow monitor capable of delivering 7,000 gpm. Courtesy of Carole Graphman.

Among Seattle Fire Department's big boats, small vessels also play an important role. Rescue Boat 5, a 25-foot Safe Boat built in 2012, is stationed with Fireboats 1 and 2. Courtesy of Rich Schneider.

Unit 99, built by Utilimaster on a 2002 International Harvester, is an indispensable element of Seattle's fireboat fleet since it carries all the tools necessary for shipboard emergencies. Courtesy of Bill Hattersley.

Inflatable Boatworks donated this Zodiac, complete with prow monitor, to the fire department in 1998. It served fifteen years before being sold. Courtesy of Last Resort Fire Department.

Tacoma

In the southern reaches of Puget Sound, Washington's port city of Tacoma has a significant nautical history that includes fireboats. Beginning in 1902, Tacoma Tug & Barge Co.'s tug *Fearless* protected the city's thirty-eight miles of waterfront that teemed with the busyness of sawmills and lumberyards. Port cities commonly relied on local tug companies for fire protection before fireboats appeared along municipal waterfronts. Built in 1900, the 80-foot *Fearless* was one of the most powerful tugs afloat. Joining it in 1914 was the Foss No. 12, which was the first tow boat built for Foss Launch Tug Co. of Tacoma. The Twelve, as it was called, was a 43-foot steam-powered tug built at Gig Harbor, Washington. It was designed so that one man could pilot the boat with its 75 hp engine, and tend lines. Foss reached an agreement with the city to hire The Twelve for a few dollars a day to fight fires. It served faithfully, alongside the *Fearless* and Foss No. 9, which had been built in 1907 as the *Violet*. The *Fearless* went to Foss in 1925 and was renamed *Foss No. 21*.

After more than a half century of work in Tacoma Harbor, The Twelve's hull was set ablaze in Commencement Bay on July 4, 1967. The monitor atop the deckhouse was added during the 1930s as part of repair work following a sinking. Besides fighting fires, *Foss No. 12* was used to wash down barges. Courtesy of Puget Sound Marine Historical Society.

Perhaps it was confidence in the trio of commercial tugs equipped with monitors that caused the Tacoma Fire Department to put off purchasing a fireboat until it neared its half-century mark. Designed by T. M. Rowlands and built by Tacoma's Coast Line Shipbuilding Co., the 96-foot fireboat was capable of 10,000 gpm delivered through seven monitors, one of which was incorporated into a stout, 26-foot tower. A massive 6,500-gpm bow monitor, designed and cast by Tacoma Fire Department personnel, was named Big Bertha. Pumps also supplied twelve discharge outlets, for which the vessel carried 2,500 feet of hose.

Tacoma's Fireboat 1 in 1970 seen from the 11th Street Bridge. The giant bow monitor, nicknamed Big Bertha, could throw a stream of water nearly 500 feet. Courtesy of Rich Schneider.

Named Fireboat No. 1, the vessel featured a steel hull and superstructure. Beneath its steel deck were five Sterling Viking II 425 hp gasoline engines, only one of which was dedicated to propulsion. The remaining four drove four De Laval, five-stage centrifugal 2,500-gpm pumps; two of the engines could also be used for propulsion.

When the *Foss No. 12* reached its golden anniversary year, its overhauls and engine changes had outlasted its wooden hull. The vessel's work in Tacoma Harbor ended in 1966, and on July 4, 1967, its hull was towed out into Commencement Bay and set afire. The following year, the *Foss No. 9* was sold to the city of Renton, Washington, as a fireboat along the southern shores of Lake Washington.

Fireboat No. 1, which had begun service on October 1, 1929, was scheduled for replacement when it reached the fifty-year mark. But its replacement would deviate dramatically from conventional fireboat design.

Research for new fireboats to meet Tacoma's needs actually began in 1973 with a study undertaken by the US Maritime Administration. In addressing the study's recommendations for fireboat design, Tacoma officials were thinking completely out of the box. Their conclusions defied logic in the eyes of stalwart traditionalists, and even bordered on the ridiculous. Tradition dictated that fireboats be steel vessels powered by mighty diesel engines driving sturdy pumps, while retaining the lines of tugs from which they had evolved. During extensive research of fireboats and fire departments, Tacoma's design team focused on surface-effect-system (SES) vessels. The concept of a rugged, yet lightweight, vessel riding on a cushion of air was nothing new. Vosper Hovermarine of Southampton, England, had pioneered the design and construction of SES passenger boats. The advantages of the design were innumerable; they were lighter, faster, more maneuverable, more fuel efficient, less costly, required less maintenance, and created less wake than planing hull fireboats. Basically, SES vessels skim across the surface of the water on a cushion of air trapped beneath the boat by flexible skirts. Downdraft from fans lifts the craft above the surface, reducing drag and allowing high speeds.

So confident were Tacoma officials that they ordered two SES fireboats, which accounted for all of the available funds allocated for new fireboats. Only Vosper Hovermarine bid on Tacoma's proposal since other boat builders deemed the concept too unorthodox and financially risky. As if going beyond US borders to order fireboats wasn't enough to rankle the skeptics, Tacoma ordered the vessels painted non-traditional lime green.

Hovercraft consultants Waterside Marine Technology engineered Tacoma's unique twins, which were Vosper's Model HM221, based on the company's HM218. Built of fiberglass, each catamaran-type boat weighs 39 tons and measures 70 feet long. Twin 450 hp Detroit diesel engines are dedicated to propulsion, driving twin screws, while a 400 hp diesel engine powers six aluminum fans to create an air cushion beneath the vessel, providing lift. When disengaged, this powerplant drives a Waterous 2,000-gpm pump, along with two other pumps driven by a 570 hp Detroit diesel, giving the vessel a rating of more than 7,000 gpm. A hydraulically operated 35-foot telescoping ladder and pipeway with 1,500-gpm monitor was incorporated into a 2,800-pound capacity crane mounted on the afterdeck. Two 2,500-gpm bow monitors, two 2,500-gpm under-pier monitors, and a 5,500-gpm main monitor all are remotely controlled.

The boats were named after local landmarks, Point Defiance and Commencement Bay. First to arrive was the *Defiance,* which began service as Fireboat 5 on December 24, 1982, followed by the *Commencement* on February 28, 1983, as Fireboat 1. Having served fifty-four years, the former Fireboat 1 was retired, and on December 2, 1983, it was added to the National Register of Historic Places. In March 1986, Tacoma's original Fireboat 1 was lifted out of the water to become a permanent display at the city's Marine Park.

Impressive service by the SES fireboats, which ranged from water rescues to battling major blazes, stilled the disbelievers. In an all-too common scenario, funding for the SES fireboats dwindled during the 1990s, forcing the *Defiance* into retirement in August 2005. It was used for parts for the *Commencement,* which eventually underwent major rebuild.

Tacoma's SES twins were America's first catamaran-style and SES fireboats. *Commencement* demonstrates its ability to skim a choppy surface at speeds in excess of 40 mph while creating minimal wake. The monitor at the tip of its elevating boom was rated at 1,500 gpm. Courtesy of Rich Schneider.

Apparent here on the *Defiance*, minus its elevating boom, are the exaggerated twisting bows of its monitors, which reduced turbulence. The massive pilothouse monitor delivered 5,500 gpm, while the two forward and two under-wharf nozzles each deliver 2,500 gpm. Courtesy of Rich Schneider.

Tacoma Fire Department's surface effect ship, the fireboat *Defiance*, with Fireboat 1, prior to beginning service in 1982. Courtesy of Rich Schneider.

Tacoma's waterways received an extra measure of safety with the arrival in 2012 of a waterjet-powered fast response fireboat. The 30-foot, 1,800-gpm high-tech vessel came from Canada's MetalCraft Marine, which critics noted was Tacoma's third fireboat built outside the US. Christened *Destiny* on May 12, 2012, the boat was berthed alongside the *Commencement* under the Murray Morgan Bridge on the Thea Foss Waterway. It was cross-staffed by members of nearby Station 6, and the following year took up station at the Tacoma Yacht Club for cross-staffing by members of Station 14.

The fireboat *Destiny* took to the waters through a partnership between the Tacoma Fire Department and the Port of Tacoma. The Firestorm 30 is rated at 1,800 gpm and can travel at speeds of more than 40 mph. Courtesy of MetalCraft Marine, Inc.

In the works for Tacoma is a new Firestorm 50 from the Canadian connection.

Given the immensity of the Puget Sound Region, fireboats from a number of communities work in cooperative effort to provide fire suppression and rescue coverage of the vast waterways. Continuing the 150-year-old legacy of waterfront protection along the coast of the Pacific Northwest, modern fireboats, by comparison to the old big boats, are very different machines. They feature wide hulls to accommodate powerful engines and pumps, and to provide low-speed maneuverability and stability. Navigation is now pinpoint, thanks to high-tech wizardry, with similar gear that allows pilots to see through smoke and foul weather. And they come equipped to function in Chemical, Biological, Radiological, Nuclear, and Explosive (CBRNE) environments. Because of such threats in the twenty-first century, federal Homeland

During a 2007 overhaul, which included engine upgrades, the *Commencement* was repainted traditional fire service colors. The large monitor delivers 5,500 gpm. Courtesy of Tacoma Fire Department.

Tacoma's Fireboat 1 is the only purpose-built fireboat in US history to have protected a major port by itself for more than half a century. In 1989, it took its place as a National Historic Landmark at Ruston Way, Tacoma. Courtesy of Madeline Houston.

Security grants have accounted for record numbers of fire-rescue boats in the Northwest.

For example: A Fort Security Grant opened the door to the Vancouver, Washington's fire department to purchase three vessels, including a 46-foot, 3,000-gpm fireboat named *Discovery*. Built by Munson Boats of Burlington, Washington, the fast-response vessel features a landing craft hull to access the shoreline of the Columbia River. The same grant cleared the way for two 1,500-gpm 30-footers built by North River Boats for Clark County and the Port of Astoria.

In the southern reaches of the Puget Sound, covering mainly Lakewood and University Place, West Pierce Fire & Rescue operates a 37-foot catamaran fire-rescue boat named *Endeavor*. The vessel was purchased from Moose Boats with a Port Security Grant. Located on Tacoma's Narrows Marina, the 3,000-gpm *Endeavor* uses twin 425 hp diesel engines coupled to dual Hamilton jet drive to reach 40 mph and to drive twin Hale pumps. The Bellingham, Washington, fire department followed suit with a contract for a 38-foot aluminum catamaran with the ability to pump 2,800 gpm while maintaining full maneuverability.

Despite Alaska's expansive land area with corresponding shoreline, weather conditions in the region often are not conducive to fireboat operations. And large boats built to withstand cold weather extremes are cost prohibitive, necessitating cooperative effort by various maritime agencies to respond to water emergencies.

Prominent among Alaska's small number of fire-rescue vessels is Ketchikan's 45-foot, all-aluminum fireboat built during 1986 by Workboats Northwest. Named after a Ketchikan Fire Department member killed in the line of duty, the *Harry Newell* replaced a 65-foot wooden fireboat, which despite its 4,000-gpm rating was deemed too slow. The *Newell's* twin 410 bhp Detroit diesels propel the vessel, and power four American pumps to provide 5,000 gpm through four monitors.

Washington's West Pierce Fire & Rescue operates this Moose Boats catamaran named *Endeavor*. Special features include a wrap-around fendering system, waterjet drive, and pumps that produce 3,000 gpm. Courtesy of West Pierce Fire & Rescue.

Built in 1986 by Workboats Northwest for rugged work with Alaska's Ketchikan Fire Department, this 45-foot, all-aluminum boat is rated at 5,000 gpm. Special features include push knees and bow thrusters. Courtesy of Murray Lundberg.

This 20-foot boat operated by Alaska's Point Baker Volunteer Fire Department is the only emergency vessel within a 50-mile radius. The Swenson boat with lowering bow ramp is powered by a 70 hp engine, with a small diesel pump to supply a monitor and hoses. Courtesy of Tessa and Kristina Jackson.

The *Newell's* main monitor delivers 2,500 gpm, while two stern monitors are rated at 750 gpm. A tow bitt and dive platform with ladder are also built into the stern. Courtesy of Murray Lundberg.

CHAPTER 9: HAWAII

Throughout the Hawaiian Islands, rescue boats, some of which are equipped with pumps for firefighting, provide ocean and waterfront safety for the state's eight islands. Most prominent among these vessels, and the state's largest, are those operated by the Honolulu Fire Department on the Island of Oahu. Waterborne fire protection for the capital city and major seaport began in 1951 with the 87-foot fireboat *Abner T. Longley*. It was built as hull number 252 in the yard of Albina Engine & Machine Works of Portland, Oregon. The only other fireboat built by Albina was the Bethel F. Gifford, delivered to Los Angeles in 1962. The company turned out 429 vessels before being sold to Dillingham Corp. in 1971, becoming Cascade General in 1986.

The *Longley's* single screw was driven by four 165 hp Detroit diesel engines, with two additional engines driving six pumps to achieve a rating of 9,000 gpm. When not powering the drive system, each propulsion engine could power a pump.

A study by the State of Hawaii Department of Transportation during the late 1980s included the option of refurbishing the *Longley,* but concluded that a new vessel was needed to protect major hazard areas in and around Honolulu, including petrochemical facilities at distant Barber's Point. Design work for the *Longley's* replacement was at the hands of Seattle's naval architects Guido Perla & Associates. The contract for building the vessel went to Moss Point Marine of Pascagoula, Mississippi, which had built Long Beach's twin fireboats four years earlier.

Built of steel, the vessel is 110 feet long, with a 22-foot beam and drafting 7 feet. Its power for both propulsion and pumping is derived from two 1,100 hp Detroit diesel engines, which gives the boat a speed of 23 mph. Two Goulds fire pumps give the vessel a rating of 7,500 gpm, which is delivered through four monitors—a 5,000-gpm Stang bow monitor, two on the cabin afterdeck, and a 2,000-gpm monitor atop a

Honolulu, Hawaii's, fireboat, *Abner T. Longley*, was delivered in March 1951. The 9,000-gpm *Longley* served for half a century before being replaced. Author's Collection.

Stang Hydrolock tower that can elevate to 50 feet above the waterline. Four underwater discharges act as thrusters to counteract nozzle reaction, in addition to 16 hose discharges about the main deck. Foam concentrate is carried below in two 500-gallon tanks, for delivery through three Skum proportioners.

The vessel arrived at Honolulu during November 1989 and was commissioned on February 15, 1990, and named *Moku Ahi,* Hawaiian for Fire Island. Built to operate offshore, the *Moku Ahi* began service at Pier 15, alongside a fire station that accommodates its crew consisting of officer, pilot, marine engineer, pump engineer, and two firefighters.

The *Moku Ahi* ushered out the *Longley*, which was retired and sold at auction in 1993. The *Longley* was then rebuilt to US Coast Guard specifications as a 149-passenger harbor cruiser.

With removal of the tower and hose reels, and the addition of railings, canopy, and a dummy funnel to present an old-fashioned look, the *Longley* was transformed into a tour boat. The three monitors were retained and used to present water displays for various events. It is seen here at Barbers Point in September 2004. Author's Collection.

The fireboat *Moku Ahi* is painted yellow in keeping with the color scheme for apparatus of the Honolulu Fire Department. Mounted on the stern ramp of the 110-foot fireboat is a 15-foot Boston Whaler. Author's Collection.

FIREBOATS DESIGNATED NATIONAL HISTORIC LANDMARKS

A small number of fireboats, all built during the first forty years of the twentieth century, have attained status as National Historic Landmarks. They are:

Edward M. Cotter—Buffalo, NY; built 1900; became NHL 1996

Duwamish—Seattle, WA; built 1909; became NHL 1989

Deluge—New Orleans, LA; built 1923; became NHL 1989

Ralph J. Scott—Los Angeles, CA; built 1925; became NHL 1989

Fireboat No. 1—Tacoma, WA; built 1929; became NHL 1989

John J. Harvey—New York, NY; built 1931; became NHL 2000

Firefighter—New York, NY; built 1938; became NHL 1989

City of Oakland/USS Hoga YT-146—Oakland, CA; built 1940; became NHL 1989

GLOSSARY OF NAUTICAL TERMS

abaft. toward the vessel's stern, also called aft
abeam. at right angles to the keel
AFFF. aqueous film-forming foam
afterbody. portion of the vessel aft of amidships, also refers to an afterhouse or afterdeck
air draft. height of the vessel above the water
amidships. in the vicinity of the vessel's middle
anchors aweigh. when the anchor is clear of the bottom
astern. toward the rear of, or behind, a ship
athwart. side to side, across the line of a vessel's course
beam. the width, or breadth, of a vessel at its widest point
bell boat. vessel operated from the engine room with the pilot using, first, manual bell pulls of coded signals, and later telegraph
berth. the place where a vessel is tied up alongside a wharf
bhp. brake horsepower (also termed shp, for shaft horsepower), the measure of an engine's horsepower before loss of power caused by drivetrain components
bilge. the lowest recess of a ship into which water drains
binnacle. pedestal near steering wheel for mounting the compass
bitt. stout metal post fastened to a deck to which towing or mooring lines are secured
bollard. upright wooden or iron post on a wharf to which mooring lines are secured
booby hatch. watertight covering over a deck opening for a stairway or ladder
bow. the front of a vessel
bridge. high transverse platform often forming the top of a bridge house, or pilothouse, where the vessel is controlled
buffalo rail. the stern portion of a bulwark
building slip. place where a ship is built before launching
bulkhead. partition walls that subdivide vessel's interior into compartments to protect against spread of water and fire
bulwark. plates that extend beyond the main deck to form railings
bunker. a compartment, often below deck, for storing coal or other fuel
button. short mushroom-shaped bitt
capstan. a revolving device vertically mounted for heaving in mooring lines, generally not equipped for chain, like a windlass
carvel-construction. planks butted to form flush seams
CBRNE. Chemical, Biological, Radiological, Nuclear, and Explosive
chine. the line formed at the point where vessel's sides and bottom meet
cleat. wood or metal device with projecting arms or horns attached for securing mooring lines, the term cavil sometimes refers to cleats of extra size and strength
coaming. a vertical barrier that protects deck openings
companion way. a deck opening with steps or ladders leading from one deck level to another
cowl ventilator. large, metal 90-degree elbow tubes on open decks, usually having bell-shaped openings to increase air intake or exhaust; could be swiveled to take advantage of wind
cradle. wooden or steel structure built up from sliding ways for supporting a vessel
cribbing. foundations of heavy blocks and timbers that support a vessel during construction or in dry dock
cuddy. small cabin or galley
davit. a crane arm for lowering and raising ships' boats and other items
deck house. superstructure built atop main deck
displacement. the weight in tons of 2,240 pounds of the vessel and its contents, also called loaded displacement
draft. the distance that a vessel's hull is beneath the surface of the water; in the fire service, the term refers to pulling water through suction hoses by creating negative pressure on the vacuum side of the pump

draft marks. numbers marked on each side of the bow and stern showing the distance from the bottom of the keel to the waterline

dry dock. a dock in which a vessel is kept out of the water for construction or repair

fantail. the overlapping stern section of vessels that have round or elliptical after-sections of decks

fast. a rope or chain used to moor a vessel to a wharf

fender. devices built into or hung over the sides of vessels to prevent rubbing against vessels or piers; permanent fenders are hardwood or steel runs secured to the hull above the waterline; suspended devices might be bundles of wood, rope, or cork, also woven cane or tires; modern fenders usually are molded rubber

fidley deck. raised deck over the engine and boiler rooms, usually surrounding the funnel

fit-out. process of preparing a vessel for service after period of inactivity

forecastle. a structure or elevated deck at a vessel's bow, or an area below the main cabin

founder. to sink due to loss of reserve buoyancy

freeboard. the vertical distance between the main deck and waterline

freeing ports. openings in bulwarks that allow water to flow off the deck

funnel. a vessel's smokestack

galley. ship's kitchen

gangway. temporarily affixed access ladder or other means of boarding a vessel

gpm. gallons per minute, a fireboat's water delivery capability by means of monitors and discharge outlets, often used to generally describe the vessel

gross tonnage. the entire internal cubic capacity of a vessel expressed in tons of 100 cubic feet to the ton

gunwhale. upper edge of a vessel's side to prevent items from being washed overboard

guy. wire, rope, or chain cables that support booms, davits, etc.

hailing port. home port of a vessel, often displayed on the vessel's transom

harbor. an area of water that provides a natural or artificial haven for vessels; differs from a port, where vessels interact with shore installations

hawser. a large rope, 5 inches or more in circumference, used for heavy work, such as towing; called a mooring line when used to secure a vessel to a dock or pier

helm. the wheel steering position, helmsman being the pilot

hull. the portion of a vessel that lies below the main deck

keel. the bottom structural centerline of a vessel, the backbone of its frame

knot. speed measurement of one nautical mile per hour, a nautical mile being 6,082.66 feet—one knot equals 1.15 mph

laid up. out of service

LNG. Liquified Natural Gas

making way. moving through the water under propulsion

manifold valves. a grouping of valves in a piping system

monitor. correct term for heavy stream appliance mounted to decks, also called water cannon, turret, deck gun, or deck pipe

paulin. canvas used as a shelter for crew or to cover deck equipment

pile. timbers driven into the bottom and projecting above the water; pilings at the end of a pier are called fender piles

pilothouse. deckhouse with steering wheel, power controls, compass, charts, etc.

port side. the left side of a vessel, facing forward

prow. the bow above the water

Purple-K. dry chemical extinguishing agent comprised mainly of potassium bicarbonate, which is highly effective in fighting flammable liquid fires; its violet color distinguishes it from other dry agents

rake. inclination from vertical of a mast, funnel, etc.

scupper. opening or drains in the side of a vessel to carry off rainwater or seawater

sea chest. portion of the hull through which water is drawn into the pump

sheer. upward curvature or sweep of the deck toward the bow or stern

side lights. running lights of a vessel underway; red at port side, green at starboard

skeg. large fin that aids directional stability

skeleton tower. a tower, usually of steel, with exposed framing members

slip. space between two piers for berthing a vessel

spuds. long poles that are lowered through wells in a vessel's hull to reach bottom, thereby securing a vessel in place

stanchions. vertical columns made of pipe or steel to support decks, rails, etc.

starboard. right side of a ship, facing forward

stem post. sturdy post built into the bow stem

stemwinder. nickname for a vessel having its house and machinery at the stern

strake. a raised, streamlined row of material on the hull running fore and aft

strongback. sturdy, long wooden beam used as a lateral support for a vessel in drydock

superstructure. upper portion of a vessel

taffrail. railings atop bulwarks

tractor tug. maneuverable tug propelled either by a Z-drive or cycloidal system rather than traditional screw propeller

trailboard. decorative boards on both sides of the bow carrying the vessel's name

transom. flat surface that forms the stern of a square-ended vessel

trunk deck. deck covering the main body of the vessel

visor. small inclined awning of canvas or metal at the top of the pilothouse

ways. timber structure on which a vessel is built or dry docked

wharf. shoreside facility extending out into deep water where ships moor for loading or unloading

wheelhouse. same as the pilothouse, called such because it housed the pilot's station with large wheel for steering; in modern high-tech vessels, often the wheels were replaced by control levers, called joysticks

windlass. powered device for lowering or hoisting anchor chains

wiper. engineer's assistant, most junior crewman in a ship's engine room

yard. a spar attached to a mast and running athwartship for attaching halyards, lights, etc.

BIBLIOGRAPHY

Books

Bloomfield, Ron. *Maritime Bay County—Images of America.* Arcadia Publishing, 2009.

Cosgrove, Bill. *Chicago's Forgotten Tragedy.* AuthorHouse, 2010.

Cowan, David. *Great Chicago Fires: Historic Blazes That Shaped a City.* Lake Claremont Press, 2001.

Ditzel, Paul. *Fireboats.* Fire Buff Historical Division of Conway Enterprises, Inc., 1989.

Lang, Steven, and Peter H. Spectre. *On the Hawser: A Tugboat Album.* Down East Books, 1980.

LeLievre, Roger. *Know Your Ships.* Marine Publishing Co., Inc., 2013.

Ley, Jim. *Milwaukee Fire Dept. Motorized Apparatus History 1907–2002.* Milwaukee Area Technical College, Nov. 2002.

Monroe, Jeffrey W., and Robert J. Stewart. *Dictionary of Maritime and Transportation Terms.* Cornell Maritime Press, 2005.

Mutza, Wayne. *American Fire Apparatus Vol. 1—Pumpers.* Squadron/Signal Publications, 2009.

Nailen, R. L., and James S. Haight. *Beertown Blazes: A Century of Milwaukee Firefighting.* Renaissance Books, 1971

Peterson Builders, Inc. *Fifty Years of Shipbuilding Excellence.* 1983.

Scheina, Robert L. *US Coast Guard Cutters & Craft of WWII.* Naval Institute Press, 1982.

Scheina, Robert L. *US Coast Guard Cutters & Craft: 1946–1990.* Naval Institute Press, 1990.

Schneider, Richard. *Seattle Fire Department—Images of America.* Arcadia Publishing, Feb. 2007.

Skalley, Michael. *Foss: Ninety Years of Towboating.* Superior, 1981.

Sprung, Stu. *The 6 House: The History of the San Diego Firehouse Museum.* San Diego Firehouse Museum, 2012.

Swinger, Pat. *Hard Work, Good People and 100,000 Boats: 50 Years of Boat-building with Zach McClendon, Jr.* Donning Company Publishers, 2009.

Articles

Badgett, John D. "City of Los Angeles Port Fire Protection in the twenty-first century." *Fire Watch.* Los Angeles Fire Department, July 2004.

Bauman, David. "USA docks old methods of fighting portside fires." *USA Today*, May 1983.

Browning, Robert M. Jr. "Captains of the Port." Coast Guard Historian's Office, Dec. 1993.

Cahill, William. "One Hundred Years of Fireboats on the Detroit River." *Telescope*, January-February 1993. Great Lakes Maritime Institute.

Daly, Jerome Wm. "Fireboat John Kendall Tested by Detroit F.D." *Fire Engineering*, February 19, 1930.

De Santis, Solange. "Seafaring Fire Engine." *Popular Mechanics*, Oct. 1983.

Efstathiou, James. "It Rolls, Floats and Saves Lives." *Jersey Journal/Hudson Dispatch*, July 1, 1991.

Gifford, Bethel F. "The Story of Fireboat 4." *The Fireman's Grapevine*, July 1961.

Johnston, W.W. Jr. "A Look at our New Underwharf Firefighting Equipment." *The Fireman's Grapevine*, July 1961.

Legeros, Mike. "Wilmington Fire Boat History," http://egeros.com/history/wilmington/.

Magner, Thomas F. "Boston Harbor Well Protected by Coast Guard Firefighters." *Fire Engineering*, May 1944.

Matier & Ross. "Procedural Delays Sink SF Fireboat Grant." *CSFA Connection*, California State Firefighters Assn., May 15, 2013.

Meek, Clarence E. "Fireboats through the Years." *Fire Engineering*, July, 1954.

Meek, Clarence E. "Log of the Nation's Fireboat Fleet." *Fire Engineering*, Sept. 1957.

Milnes, Robert B. "Firefighting Amphibian." *Firehouse*, January 1983.

Parsons, H. *American Fire-Boats.* 1-1-1896.

Ryan, Richard A. "Old Fireboats Never Die, They Just Keep Pumping." *The Detroit News*, April 24, 1966.

Trojanowicz, Al. *Fireboats, New York Fireboats, 1988-2006.*
Tullis, George. "The Fireboat Fleet." *US Coast Guard Magazine*, January 1943.
Wade, Herbert T. "High-Pressure Fire Protection." *The American Review of Reviews*, July–December 1908.
West, Charles C. "Centrifugal-Pump Fire-Boats." *Engineering News*, December 3, 1908.
Woodard, Clarence and McCall, Walter. "Detroit's Fire Boats." *Detroit Firemen's Yearbook*, nd.

Non-author-accredited Articles

"Amphibious Fireboat for Miami." *Marine Engineering Log*, April 1983.
"Detroit Fireboat John Kendall." *Marine Engineering*, June 1930.
"Fighting Fires on Galveston's Water Front." *The Galveston Daily News*, September 28, 1913.
"New Fireboat for New York City - *The New Yorker*." *Scientific American*, March 7, 1891.
"New Fireboat is Christened." *Port of Houston Magazine*, May 1974.
"New Galveston Fireboat is Launched." *The Galveston Daily News*, February 17, 1929.
"New York Harbor Police." *Harpers*, Volume XLV. 1872.
"Pacific Coast Develops Gasoline Type Fire Boat." *Fire Engineering*, Aug. 22, 1928.
"Portland Steps Up Fireboat Potential." *Fireman*, March 1962.
"Textron Marine Systems Wins $6.5-Million Contract to Build Two SES Fireboats." *Maritime Reporter*, July 1990.
"The City's New Fire Boat." *The New York Times*, April 24, 1898.
"The 1886 Department of Docks Headquarters-Pier A." *Daytonian in Manhattan*, Tuesday, January 24, 2012.
"The New Boston Fire Boat Engine 47." *International Marine Engineering* vol. 15, June 1910.
"Wheeled Amphibians: Engineering Design Handbook." *Headquarters US Army Material Command*, January 1971.

Miscellaneous References

Delgado, James P. National Register of Historic Places Registration Form "Deluge." National Park Service, December 28, 1988.
Historical Collection of the Great Lakes. Bowling Green State University, www.bgsu.edu.
Interview of former San Diego Chief Engineer Louis Almgren by Edgar Hastings on April 17, 1957.
McFarland, John. Email to author August 2015.
Navy Seagoing Tugs and Related Craft. Bureau of Ships Navy Department, June 12, 1944.
Pennison, John L. Email to author September 2015.
31st Annual Report of the Fire Department of the City of Los Angeles, California.
39th Annual Report of the Fire Department of the City of Los Angeles, California.

ABOUT THE AUTHOR

Wayne Mutza's retirement from the Milwaukee Fire Department took him down new roads, one of which led to writing. Finding that little has been written about the fascinating history of fireboats in America, Mutza combined his enthusiasm for the fire service and his interest in work boats to produce this comprehensive volume. A historian and literary and documentary film consultant, he has published more than two dozen books and numerous articles.